AF885068

INSTRUCTOR'S MANUAL TO ACCOMPANY

Elementary Geometry

Third Edition

R. David Gustafson
Peter D. Frisk
Rock Valley College
Rockford, Illinois

PREPARED BY
DIANE R. KOENIG
Rock Valley College
Rockford, Illinois

WILEY

John Wiley & Sons, Inc. New York Chichester Brisbane Toronto Singapore

ISBN 0-471-53558-3

10 9 8 7 6 5 4 3 2

To the instructor:

This Instructor's Manual is intended to supplement ELEMENTARY GEOMETRY, third edition, by R. David Gustafson and Peter D. Frisk. It includes additional exercises, computer exercises using logo, three forms of sample chapter tests for each chapter and a two-hour final examination, and solutions to the even numbered proofs from the text.

The chapter tests are designed to take 40 minutes. If there is not adequate time to give a test at the end of each chapter, these sample tests can be combined in any way the instructor feels is appropriate. Answers to the chapter tests are found on pages 111-118 of this manual.

Diane R. Koenig

CONTENTS

CHAPTER 1

FUNDAMENTAL CONCEPTS

SUPPLEMENTARY EXERCISES

1. To determine if the product of an even integer and an odd integer is always an even integer, Jeremy finds the following products:

$$4 \cdot 3 = 12$$

$$2 \cdot 5 = 10$$

$$6 \cdot 7 = 42$$

After observing that each product is an even integer, he concludes that the product of an even integer and an odd integer is always an even integer. Is Jeremy's conclusion correct? Did he reason inductively or deductively? (Yes. Inductively.)

2. If the length of each side of an equilateral triangle with perimeter of 18 in. is tripled, what is the new perimeter? (54 in.)

3. Find the sum of the complement of the supplement of 131° and the supplement of the complement of 49°. (180°)

4. Two angles measuring $(4x)^\circ$ and $(3x + 40)^\circ$ are supplementary. Find x. $(x = 20)$

5. Attempt to construct a triangle with two right angles. What problem confronts you?

6. Construct an angle of $22\frac{1}{2}^\circ$.

7. Draw a straight angle and bisect it.

8. Draw an angle and contruct an angle that is supplementary to it. Copy the supplementary angle.

9. Draw an obtuse angle on a piece of paper. Construct another angle that is congruent to and adjacent to the first angle.

10. One of two supplementary angles is 4 times as large as the other. What is the complement of the smaller angle? (54°)

11. Prove: If two angles are complementary and equal, they are 45° angles.

12. Prove: If $\angle 1 \cong \angle 2$, then $\angle 1$ and $\angle 3$ are supplementary.

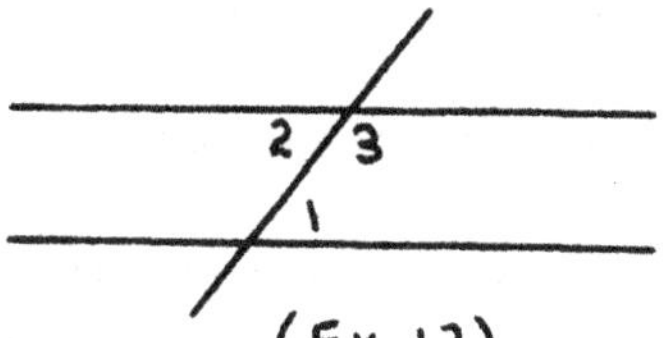

(Ex. 12)

Chapter 1 Test

Form 1

Decide if each of the following statements is true or is false. Then mark T or F in the space provided.

_________ 1. If we form generalizations from specific cases, we are reasoning inductively.

_________ 2. Each word must be defined when using deductive reasoning.

_________ 3. A theorem is a statement that has been proved.

_________ 4. A scalene triangle has a pair of congruent sides.

_________ 5. An angle of 135° is an obtuse angle.

_________ 6. An angle of 1° is 1/360th of a complete revolution.

_________ 7. A right triangle can have two right angles.

_________ 8. A triangle has three altitudes and three medians.

_________ 9. There can be only one perpendicular to a line through a given point on the line.

_________ 10. Q is the vertex of $\angle PQR$.

_________ 11. $\angle PQR$ and $\angle QRS$, $P \neq S$, must be adjacent angles.

_________ 12. Angles of 30°, 40°, and 20° are complementary angles.

_________ 13. Angles of 120° and 60° are supplementary.

_________ 14. If angles are congruent, they are vertical angles.

_________ 15. Adjacent angles can be complementary.

Solve each problem and place the answer in the space provided.

_________ 16. Solve the equation $2x + 3 = 9$.

_________ 17. Solve the equation $\frac{8x - 4}{2} = 6$.

_________ 18. Find the supplement of an angle of 57°.

_________ 19. Find the complement of the supplement of an angle of 93°.

_________ 20. Find the perimeter of an isosceles triangle that has sides of 4 cm and a base of 5 cm.

_________ 21. Find the perimeter of an equilateral triangle with a side of 6 inches.

__________22. If the perimeter of an isosceles triangle is equal to 16 m and the base is 4 m, how long is one of the equal sides?

__________23. Find the supplement of the supplement of an angle of 57°.

__________24. One of the equal sides of an isosceles triangle is 13 inches less than the perimeter of the triangle. The base is 5 inches. Find the length of one of the equal sides.

__________25. If $m(\angle ABC) = 56°$ and $\overline{BD}$ bisects $\angle ABC$, what is the complement of $\angle ABD$?

26. Use a compass and a straightedge to construct the perpendicular bisector of $\overline{AB}$.

27. Use a compass and a straightedge to construct the bisector of $\angle ABC$.

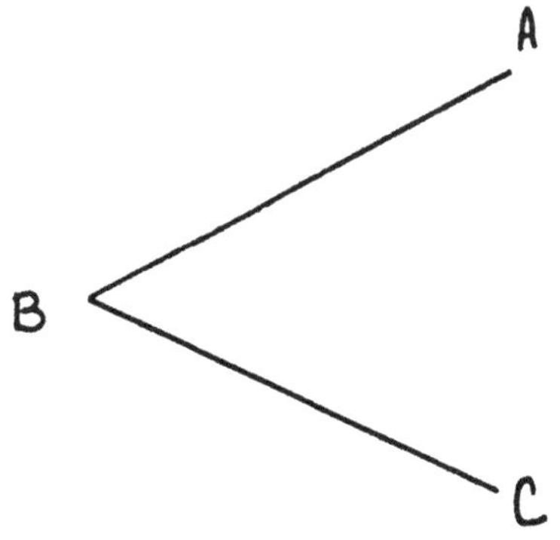

Chapter 1 Test

Form 2

Decide if each of the following statements is true or is false. Then mark T or F in the space provided.

_________ 1. If we form generalizations from specific cases, we are reasoning deductively.

_________ 2. All terms are defined when using deductive reasoning.

_________ 3. Postulates are statements that are assumed true.

_________ 4. An isosceles triangle has a pair of congruent sides.

_________ 5. A line segment has only one perpendicular bisector.

_________ 6. An angle of 91° is an acute angle.

_________ 7. An angle of 27° has no supplement.

_________ 8. An obtuse triangle has only 2 altitudes.

_________ 9. R is the vertex of $\angle QRS$.

_________ 10. Angles of $(3x + 50)^\circ$ and $(40 - 3x)^\circ$ are complementary.

_________ 11. If angles are vertical, they are congruent.

_________ 12. A right triangle is also an acute triangle.

_________ 13. A right triangle can be isosceles.

_________ 14. Vertical angles can be adjacent angles.

_________ 15. Adjacent angles whose noncommon sides form a line are supplementary angles.

Solve each problem and place the answer in the space provided.

_________ 16. Solve the equation $5x - 17 = 18$.

_________ 17. Solve the equation $\frac{2x+5}{3} = 15$.

_________ 18. Find the complement an angle of 57°.

_________ 19. Find the supplement of the complement of an angle of 47°.

_________ 20. Find the perimeter of an isosceles triangle that has sides of 8 cm and a base of 12 cm.

_________ 21. Find the length of the side of an equilateral triangle whose perimeter is 24 m.

________ 22. The perimeter of an isosceles triangle with two sides measuring 15 cm is 37 cm. Find the length of the base.

________ 23. Two angles with measures of $(5x)^\circ$ and $(x+60)^\circ$ are supplementary. Find x.

________ 24. The length of the side of an equilateral triangle is 14 inches less than the perimeter. Find the lengths of the sides.

________ 25. If $m(\angle ABC) = 76^\circ$, what is the measure of one of the angles formed when its supplement is bisected?

26. Use a compass and a straightedge to construct a line perpendicular to $\overline{AB}$, through point P.

27. Use a compass and a straightedge to copy $\angle ABC$.

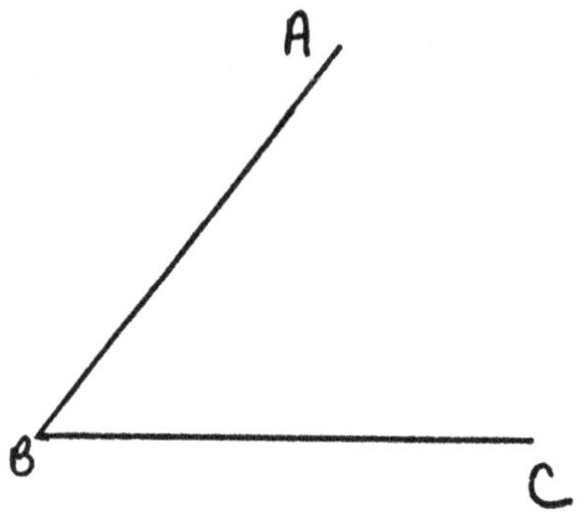

Chapter 1 Test

Form 3

Decide if each of the following statements is true or is false. Then mark T or F in the space provided.

_________ 1. Deductive reasoning contains the four components: undefined terms, defined terms, postulates, and theorems.

_________ 2. Theorems are statements that are assumed true.

_________ 3. In a construction problem, the only tools allowed are a compass, pencil, and a ruler.

_________ 4. An equilateral triangle is also an isosceles triangle.

_________ 5. An angle of 87° is an acute angle.

_________ 6. An angle of 97° has no complement.

_________ 7. An acute triangle can have a right angle.

_________ 8. In a triangle, a median can also be an altitude.

_________ 9. An angle has only one angle bisector.

_________10. There can be more than one line perpendicular to a line through a given point not on the line.

_________11. Angles of $(5x + 35)^\circ$ and $(155 - 5x)^\circ$ are supplemenetary.

_________12. Vertical angles can be supplementary.

_________13. Adjacent angles can be complementary.

_________14. A bisector of a right angle divides the angle into two complementary angles.

_________15. All pairs of nonadjacent angles are vertical angles.

Solve each problem and place the answer in the space provided.

_________16. Solve the equation $3x + 12 = 45$.

_________17. Solve the equation $\frac{2}{3}x + 4 = 10$.

_________18. Find the complement an angle of 24°.

_________19. Find the complement of the supplement of an angle of 125°.

_________20. Find the perimeter of an equilateral triangle that has sides measuring 5 cm.

_________21. The perimeter of an isosceles triangle with two sides measuring 10 meters is 25 meters. Find the length of the base.

________ 22. The second longest side of a scalene triangle is 3 cm longer than the shortest side. The longest side has measure of 10 cm. If the perimeter is 17cm, find the length of the shortest side.

________ 23. Two angles with measures of $(3x)^\circ$ and $(x+50)^\circ$ are complementary. Find x.

________ 24. The $m(\angle 1)$ is 8 times its supplement. Find $m(\angle 1)$.

________ 25. If $m(\angle ABC) = 24^\circ$ and $\overline{BD}$ bisects $\angle ABC$, what is the supplement of $\angle ABD$?

26. Use a compass and a straightedge to construct a line perpendicular to $\overline{AB}$, through point P.

27. Use a compass and a straightedge to construct the altitude from vertex A in $\triangle ABC$.

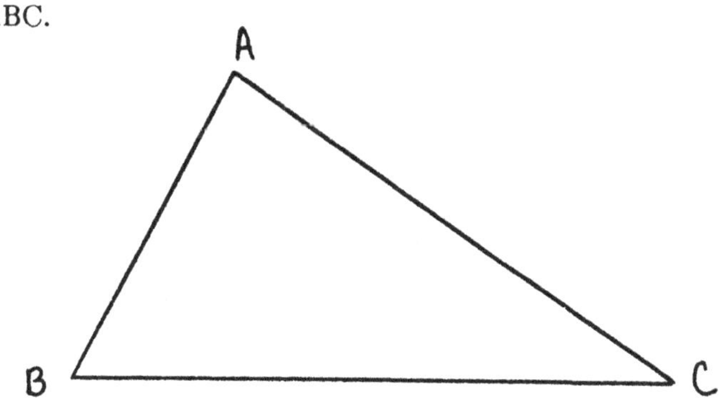

CHAPTER 2

CONGRUENT TRIANGLES AND BASIC THEOREMS

SUPPLEMENTARY EXERCISES

1. Two people stand on level ground the same distance from the base of a vertical flagpole. Prove that their distances from the top of the flagpole are equal.

2. Given: Isosceles $\triangle ABC$, with $\overline{AC} \cong \overline{BC}$.
   Isosceles $\triangle ABD$, with base $\overline{AB}$.

   Prove: $\triangle ADC \cong \triangle BDC$.

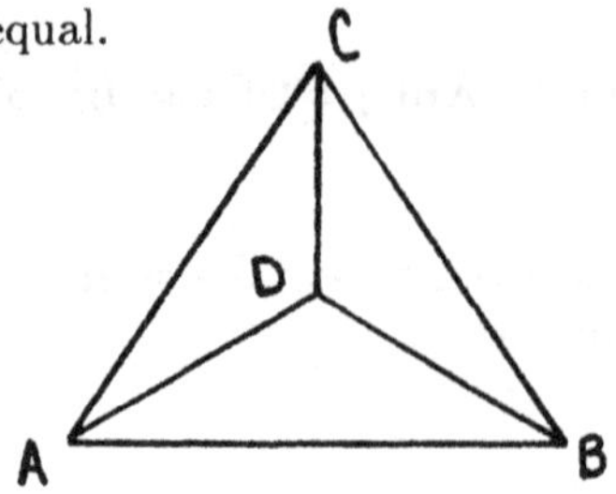

3. Given: Figure ABCD with E the midpoint of $\overline{AB}$.
   $\overline{AD} \cong \overline{BC}$.
   $\angle A \cong \angle B$.

   Prove: $\angle 1 \cong \angle 2$.

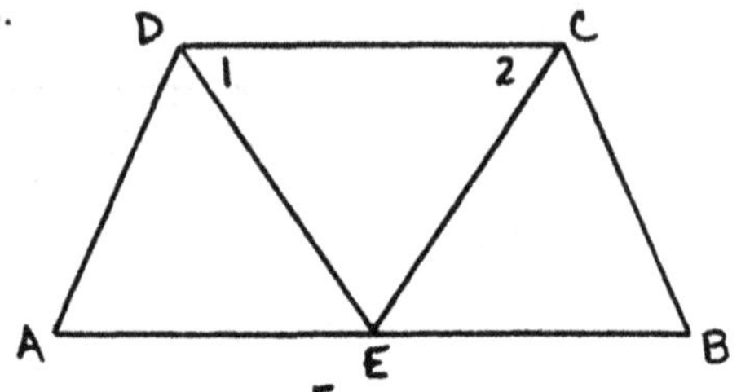

4. Given: $\triangle ABF$, $\triangle FGE$, $\triangle BCE$ and $\triangle CDG$ are isosceles with bases $\overline{AB}$, $\overline{FG}$, $\overline{BC}$ and $\overline{CD}$ respectively.

   Prove: $\triangle ADH$ is isosceles.

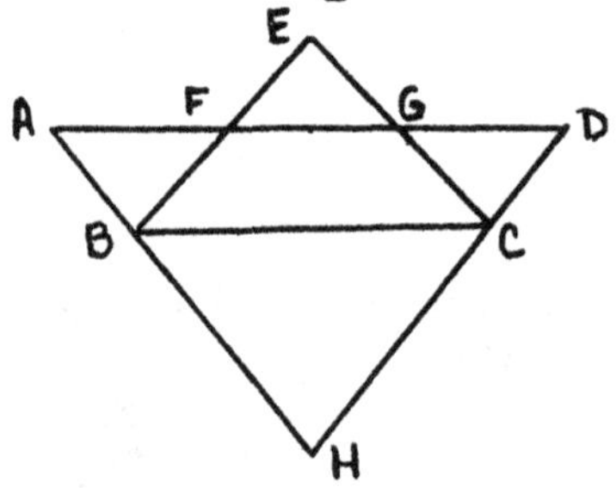

5. Given: $\overline{AD}$ with $\overline{AF} \cong \overline{FG} \cong \overline{GD}$.
   F the midpoint of $\overline{BE}$.
   G the midpoint of $\overline{CE}$.
   $\overline{AE} \cong \overline{DE}$.

   Prove: $\overline{AB} \cong \overline{DC}$.

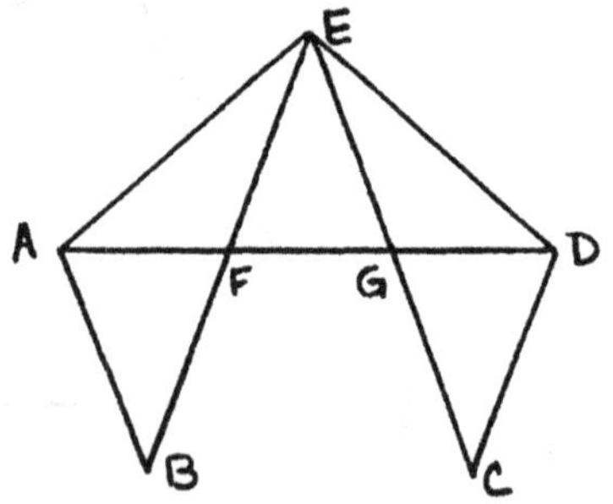

6. Given: $\angle 1$ is complementary to $\angle 4$.
   $\angle 2$ is complementary to $\angle 3$.
   $\overline{AC}$ bisects $\angle DAB$.

   Prove: $\overline{CA}$ bisects $\angle DCB$.

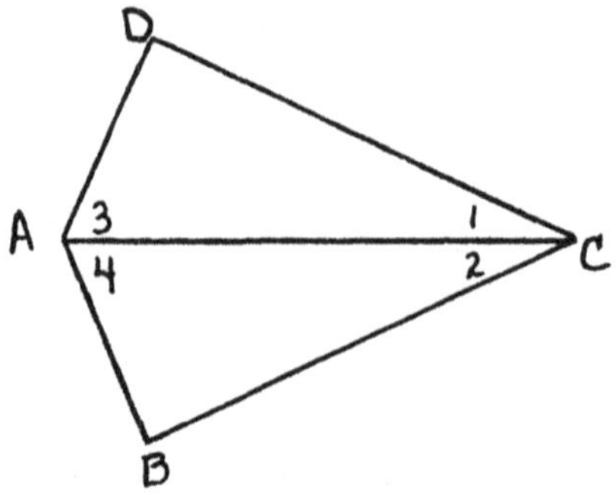

7. Given: $\angle E \cong \angle CBD$.

$\angle ABE \cong \angle C$.

Prove: $\angle E \cong \angle C$.

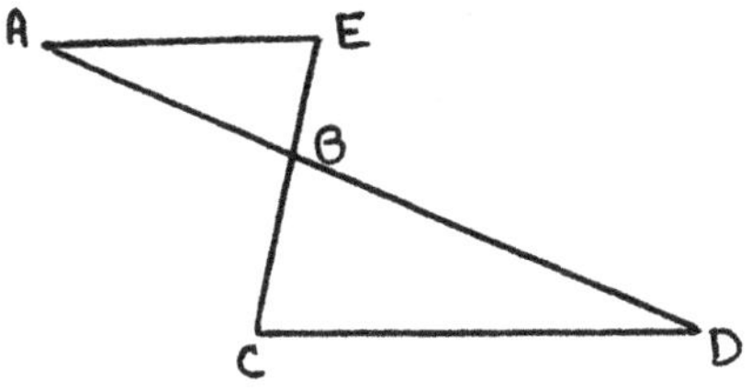

8. Prove that an equiangular triangle is equilateral.

9. Given these three sides, construct triangle ABC.

10. Given side $\overline{AB}$ and angles A and B, construct $\triangle ABC$.

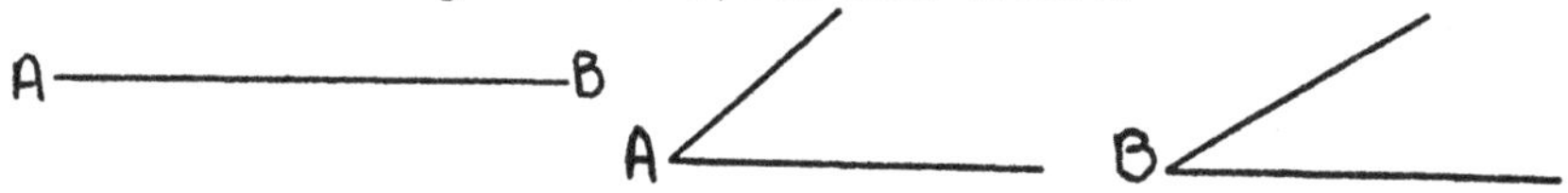

11. Given $\angle A$ and sides $\overline{AB}$ and $\overline{BC}$ construct $\triangle ABC$. What is strange about this construction? (Answer is not unique.)

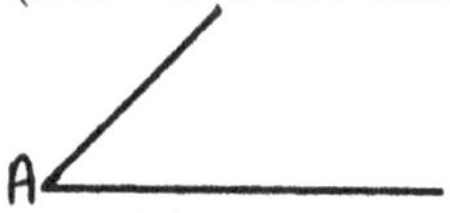

12. Given $\angle A$ and sides $\overline{AB}$ and $\overline{BC}$ construct $\triangle ABC$. What is strange about this construction? (The construction is impossible because the triangle will not "close up".)

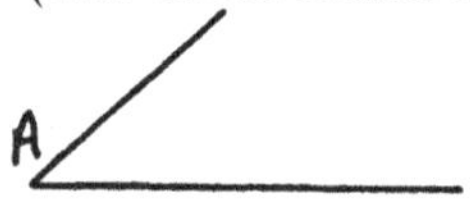

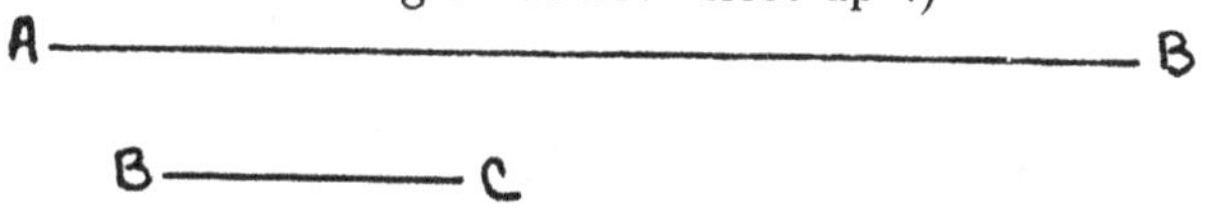

13. Construct a triangle with the given sides $\overline{AB}$ and $\overline{AC}$, and the given altitude drawn from $\angle A$ to the third side. (Solution: At point D on a line l construct a perpendicular p. Lay off altitude $\overline{AD}$ on p. From $\angle A$ describe arcs intersecting l with radii $\overline{AB}$ and $\overline{AC}$. There are two possible triangles.)

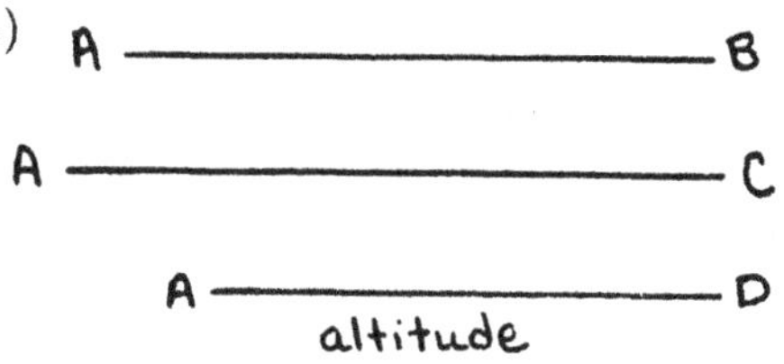

14. Given an acute angle and a line segment equal to the sum of the legs of a right triangle, construct the right triangle. (Solution: Construct $\overline{AB}$ equal to the sum of the legs and construct the given angle at A. At B, construct a 45° angle, determining point C. Drop the perpendicular to $\overline{AB}$ from C. $\triangle ADC$ is the required triangle.)

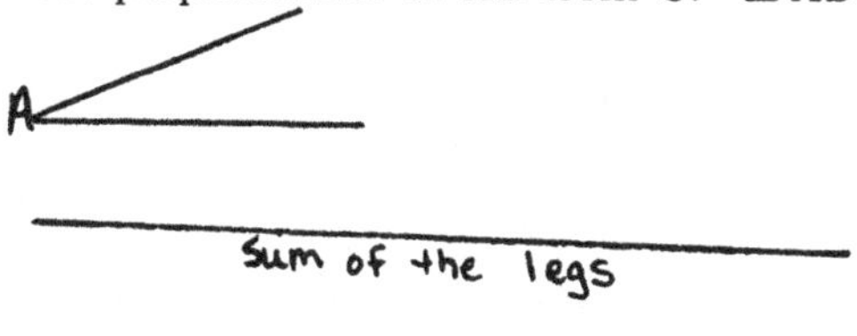

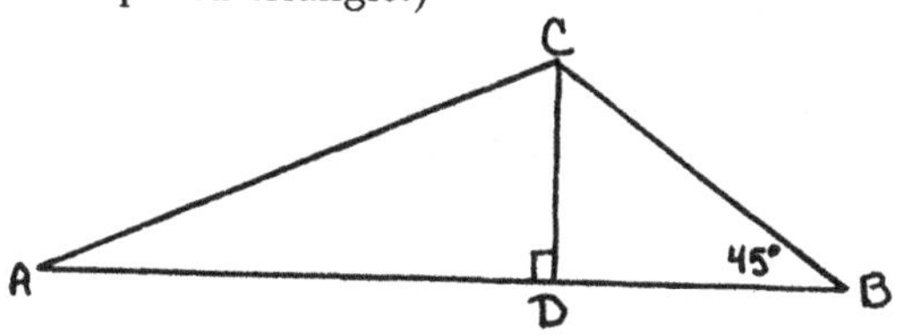

Chapter 2 Test

Form 1

Decide if each of the following statements is true or is false. Then mark T or F in the space provided.

_________ 1. If three sides of one triangle are congruent to three sides of a second triangle, the triangles are congruent.

_________ 2. If two triangles are congruent, they must be the same size and the same shape.

_________ 3. If in $\triangle ABC$ and $\triangle DEF$, $\angle A \cong \angle D$, $\angle B \cong \angle E$ and $\angle C \cong \angle F$, then $\triangle ABC \cong \triangle DEF$.

_________ 4. A triangle with three congruent angles is called an equiangular triangle.

_________ 5. The converse of a true conditional statement is always true.

_________ 6. If a triangle has two equal angles, it is isosceles.

_________ 7. The bisector of the vertex angle of an isosceles triangle bisects the base of the triangle.

_________ 8. The base angles of an isosceles triangle are congruent.

_________ 9. The side opposite the right angle in a right triangle is called the *hypothesis* of the right triangle.

_________10. Three pieces of information must be known before the congruence of triangles can be shown.

_________11. It is possible for two scalene triangles to be congruent.

_________12. Two right triangles are congruent if they have equal hypotenuses.

_________13. Equilateral triangles are not always congruent.

_________14. The symbol for congruence is *cpctc*.

_________15. Vertical angles are adjacent angles.

Solve each problem and place the answer in the space provided.

_________16. $m(\angle A) + m(\angle B) + m(\angle C) = 180°$, $m(\angle C) = 80°$ and $\overline{AB} \cong \overline{BC}$. Find $m(\angle A)$.

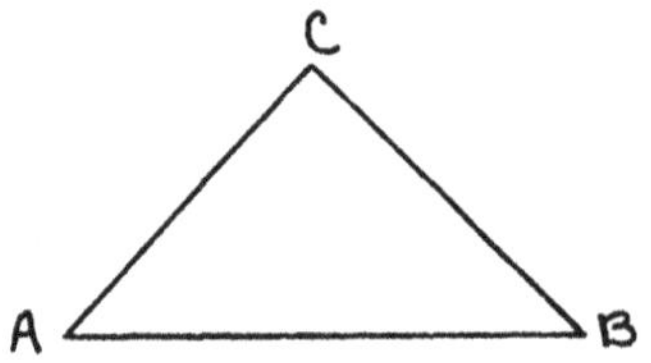

_________17. $m(\overline{AC}) = m(\overline{BC}) = 10$ in. The perimeter of $\triangle ABC$ is 28 in. Find the $m(\overline{AD})$.

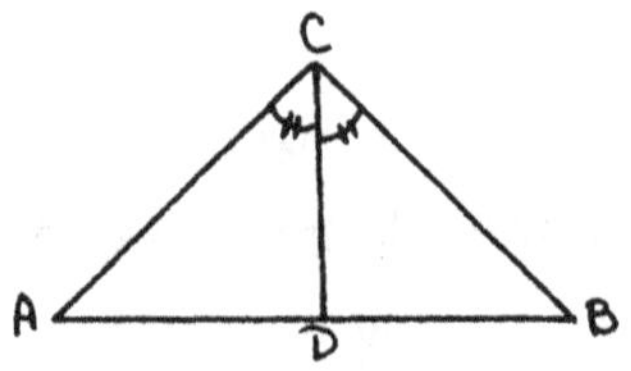

_________18. $\angle A \cong \angle E$, $\overline{AC} \cong \overline{EF}$, and $\angle C \cong \angle F$. If $m(\angle E) = 60°$, $m(\angle F) = 80°$, and $m(\angle D) = 40°$, find $m(\angle B)$.

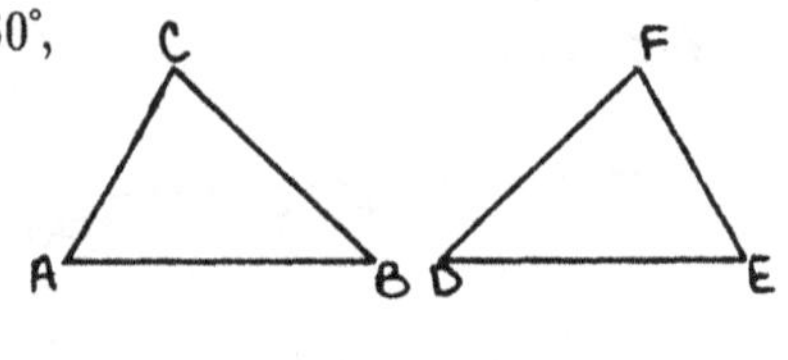

_________19. $\overline{AC} \cong \overline{BC}$, $\overline{AD} \cong \overline{BD}$ and $\overline{AE} \cong \overline{DE}$.
If $m(\overline{AB}) = 11$ m, find $m(\overline{DE})$.

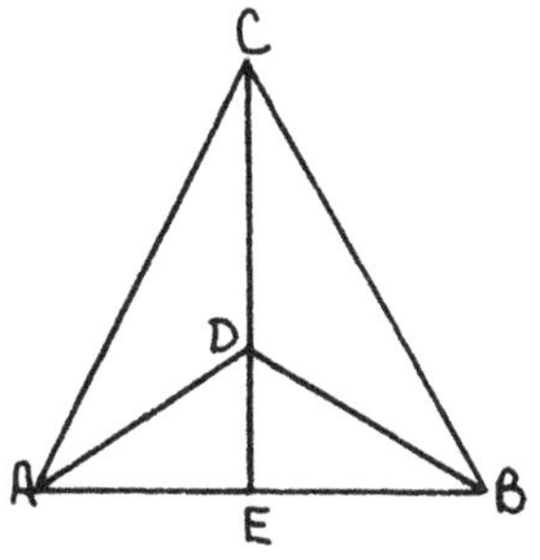

_________20. $\triangle$ ABC and $\triangle$ DEF are equilateral.
If $\overline{AC} \cong \overline{DF}$, what is the perimeter of $\triangle$ DEF given that $m(\overline{AB}) = 7$ cm?

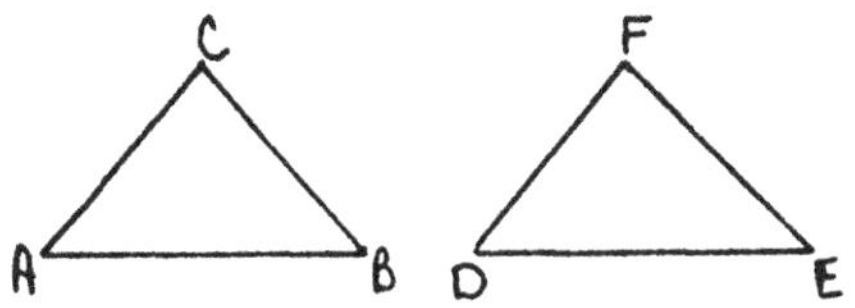

21. Given: $\angle 1 \cong \angle 2$.

$\angle 3 \cong \angle 4$.

Prove: $\overline{AC} \cong \overline{BC}$.

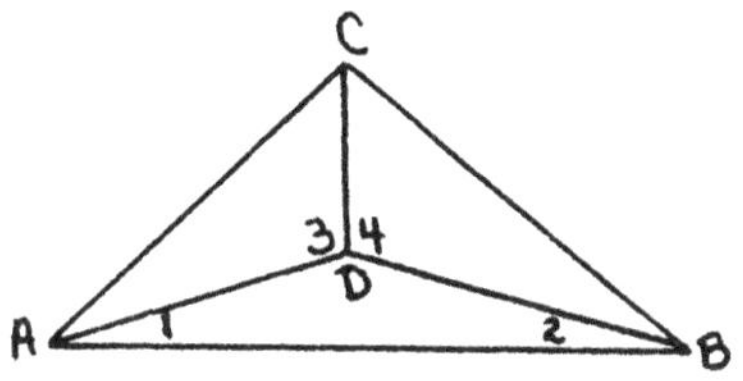

STATEMENTS	REASONS

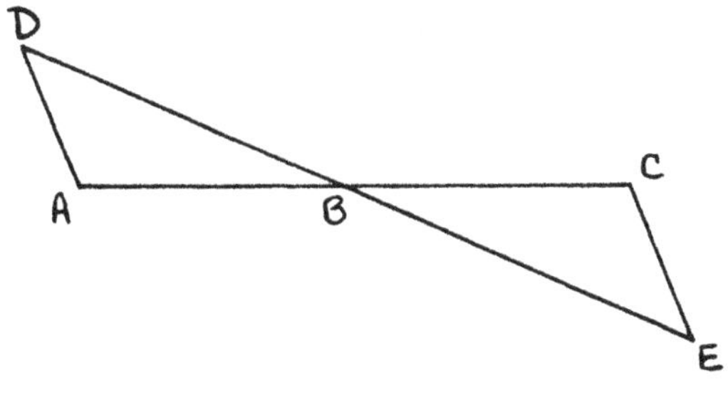

22. Given: Lines ABC and DBE bisect each other at B.

Prove: $\angle A \cong \angle C$.

STATEMENTS	REASONS

Chapter 2 Test

Form 2

Decide if each of the following statements is true or is false. Then mark T or F in the space provided.

________ 1. If two sides and an included angle of one triangle are congruent to two sides and an included angle of a second triangle, the triangles are congruent.

________ 2. If two sides and an angle of one triangle are congruent to two sides and an angle of a second triangle, the triangles are congruent.

________ 3. If in $\triangle ABC$ and $\triangle DEF$, $\angle A \cong \angle D$, $\overline{AB} \cong \overline{DE}$, and $\angle B \cong \angle E$, then $\triangle ABC \cong \triangle DEF$.

________ 4. All equiangular triangles are congruent.

________ 5. An *hypothesis* is the longest side of a right triangle.

________ 6. The bisector of the vertex angle of an isosceles triangle is perpendicular to the base.

________ 7. If an angle in a triangle appears to be a right angle, the triangle may be assumed to be a right triangle.

________ 8. If two angles in a triangle are complementary, the triangle is a right triangle.

________ 9. If two legs of one right triangle are congruent to two legs of a second right triangle, the triangles are congruent.

________ 10. The converse of a conditional statement is never true.

________ 11. It is possible for two obtuse triangles to be congruent.

________ 12. An equilateral triangle is also an isosceles triangle.

________ 13. The measures of the angles of an equilateral triangle are 60°.

________ 14. Vertical angles can be adjacent.

________ 15. In an isosceles triangle, the altitude drawn from the vertex angle is also the median.

Solve each problem and place the answer in the space provided.

________ 16. $m(\angle A) + m(\angle B) + m(\angle C) = 180°$, $\overline{AB} \cong \overline{AC}$ and $m(\angle C) = 40°$. Find $m(\angle A)$.

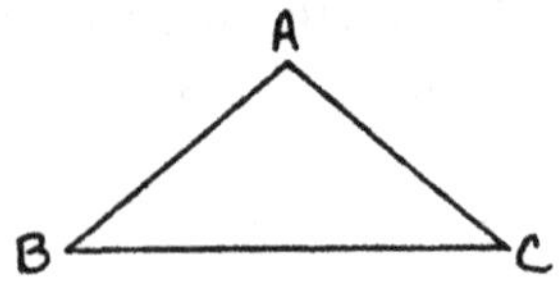

________ 17. $m(\angle A) + m(\angle B) + m(\angle C) = 180°$, $m(\angle A) = 90°$, $m(\angle B) = 45°$, and $m(\overline{AB}) = 10$ cm. Find $m(\overline{AC})$.

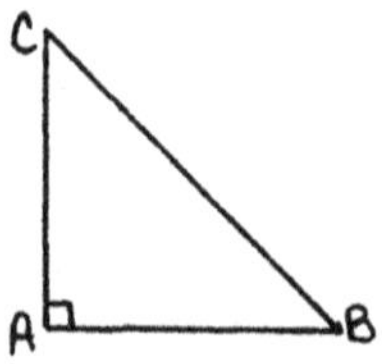

________ 18. $\angle A \cong \angle D$, $\overline{AC} \cong \overline{DF}$, and $\overline{AB} \cong \overline{DE}$. If $m(\angle D) = 20°$ and $m(\angle F) = 30°$, find $m(\angle B)$.

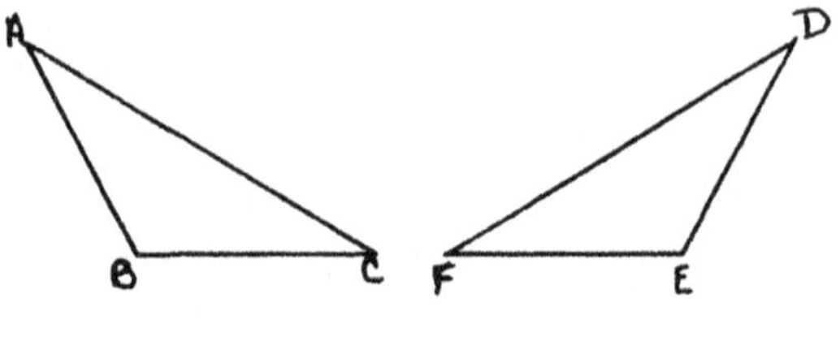

________19. From the information in the figure to the right, find the value of x.

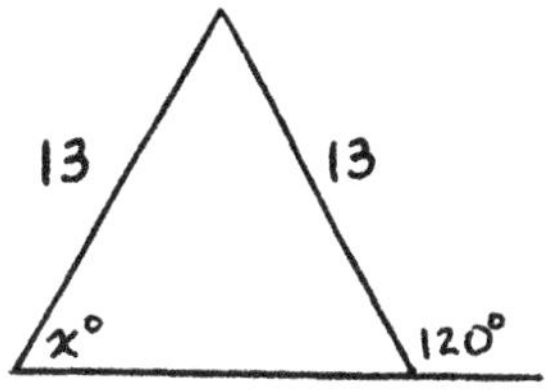

________20. $\triangle$ ABC and $\triangle$ DEF are congruent and equiangular. What is the perimeter of $\triangle$ DEF is m($\overline{AB}$) = 10 m.

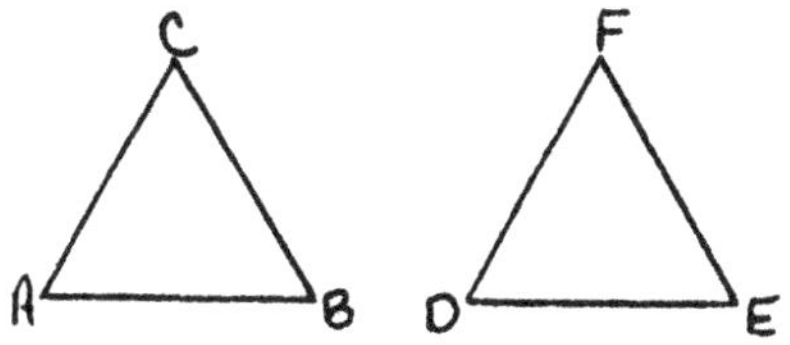

21. Given: $\overline{AE} \perp \overline{ED}$.

$\overline{CD} \perp \overline{ED}$.

$\angle AEB \cong \angle CDB$.

Prove: $\triangle$ EBD is isosceles.

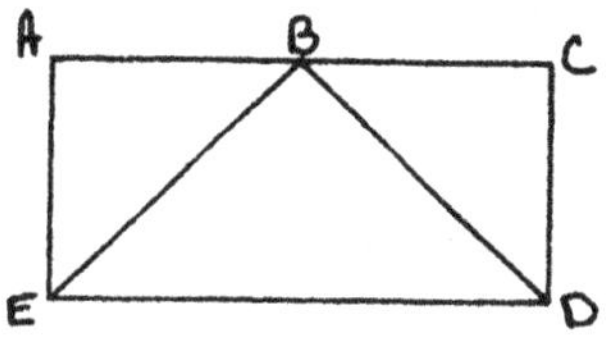

STATEMENTS REASONS

22. Given: $\angle 1 \cong \angle 4$.

$\overline{AB} \cong \overline{DC}$.

$\overline{EA} \cong \overline{ED}$.

Prove: $\overline{EB} \cong \overline{EC}$.

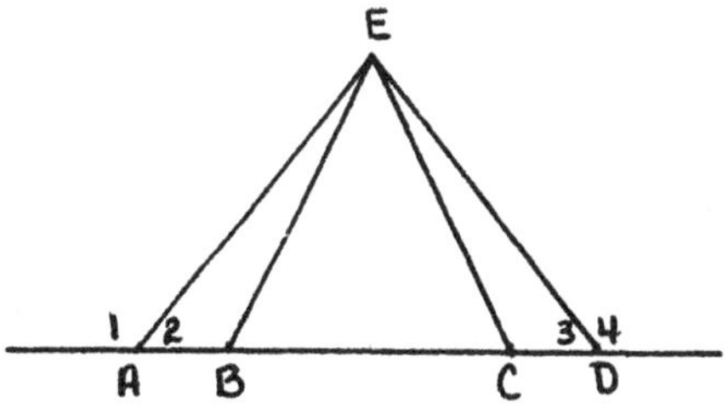

STATEMENTS REASONS

Chapter 2 Test

Form 3

Decide if each of the following statements is true or is false. Then mark T or F in the space provided.

_________ 1. If all three angles of one triangle are congruent to all three angles of a second triangle, the triangles are congruent.

_________ 2. If two angles and an included side of one triangle are congruent to two angles and an included side of a second triangle, the triangles are congruent.

_________ 3. If in $\triangle ABC$ and $\triangle DEF$, $\angle A \cong \angle D$, $\overline{AB} \cong \overline{DE}$, and $\overline{AC} \cong \overline{DF}$, then $\triangle ABC \cong \triangle DEF$.

_________ 4. All equilateral triangles are congruent.

_________ 5. The symmetric law states that if $\overline{AB} \cong \overline{CD}$, and $\overline{CD} \cong \overline{EF}$, then $\overline{AB} \cong \overline{EF}$,

_________ 6. The bisector of the vertex angle of an isosceles triangle bisects the base.

_________ 7. Diagrams are drawn so that anything that appears to be true can be assumed true.

_________ 8. If two angles in a triangle are supplementary, then the triangle is a right triangle.

_________ 9. If all three angles of a triangle are congruent, then the triangle is isosceles.

_________ 10. If a leg of one right triangle is congruent to a leg of a second right triangle, then the triangles are congruent.

_________ 11. An equilateral triangle is also isosceles.

_________ 12. The *hypothesis* is the clause following the word If in an If . . . then . . . sentence.

_________ 13. The converse of a conditional statement is always true.

_________ 14. If two triangles have the same shape, they are called congruent triangles.

_________ 15. One valid way of showing congruence is ASA $\cong$ ASA.

Solve each problem and place the answer in the space provided.

_________ 16. $m(\angle A) + m(\angle B) + m(\angle C) = 180°$, $\overline{AB} \cong \overline{AC}$ and $m(\angle A) = 60°$. Find $m(\angle C)$.

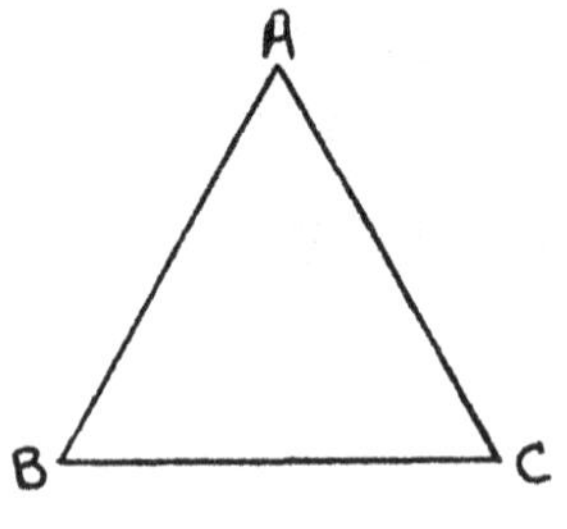

_________ 17. $m(\angle A) + m(\angle B) + m(\angle C) = 180°$, $\overline{CD}$ is the perpendicular bisector of $\overline{AB}$ passing through $\angle C$, and $m(\angle A) = 30°$. Find $m(\angle ACD)$.

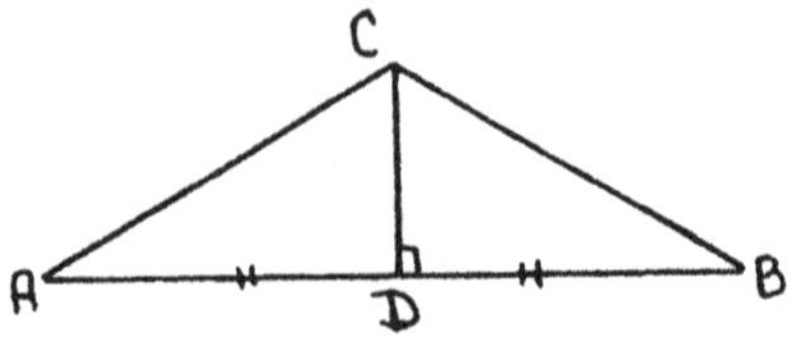

________18. $\angle A \cong \angle D$, $\angle B \cong \angle E$, and $\overline{AB} \cong \overline{DE}$. If $m(\angle B) = 50°$, and $m(\angle F) = 60°$, find $m(\angle A)$.

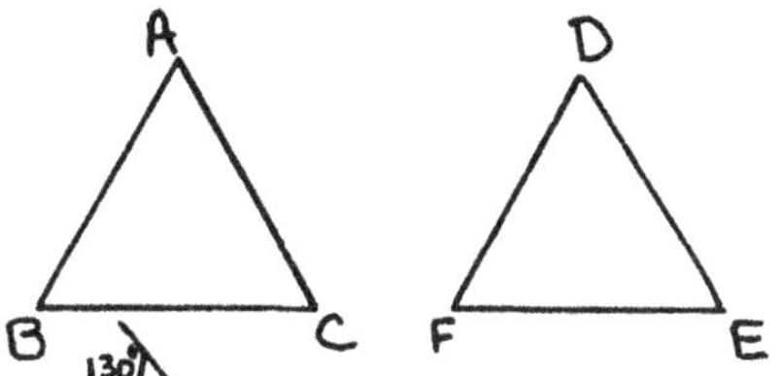

________19. From the information in the figure to the right, find the value of x.

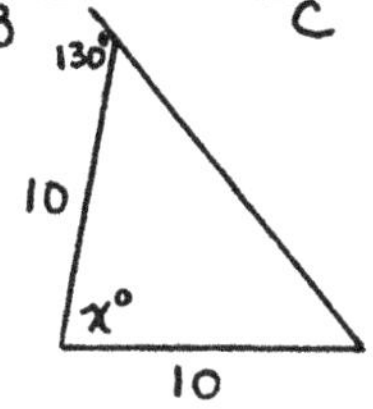

________20. $\triangle ABC \cong \triangle DEF$. Find the perimeter of $\triangle ABC$.

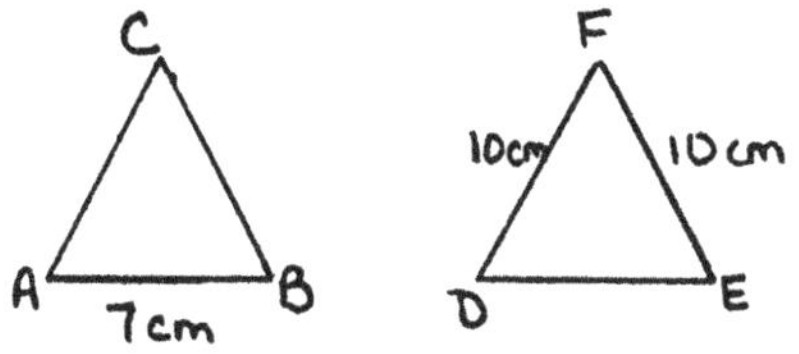

21. Given: $\triangle ABE$.

$\angle A \cong \angle B$.

$\overline{AC} \cong \overline{BD}$.

Prove: $\angle 1 \cong \angle 2$.

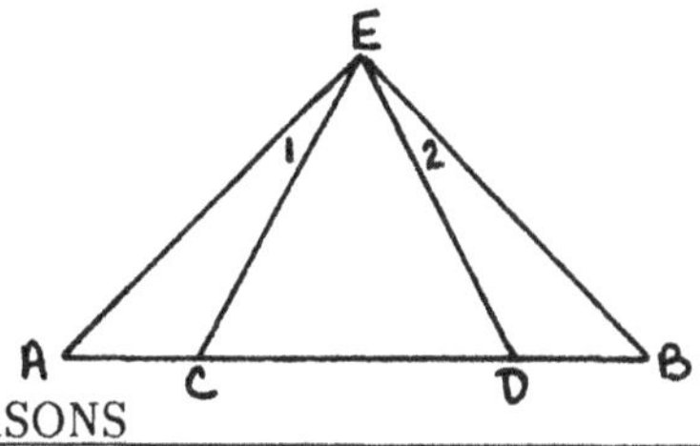

STATEMENTS	REASONS

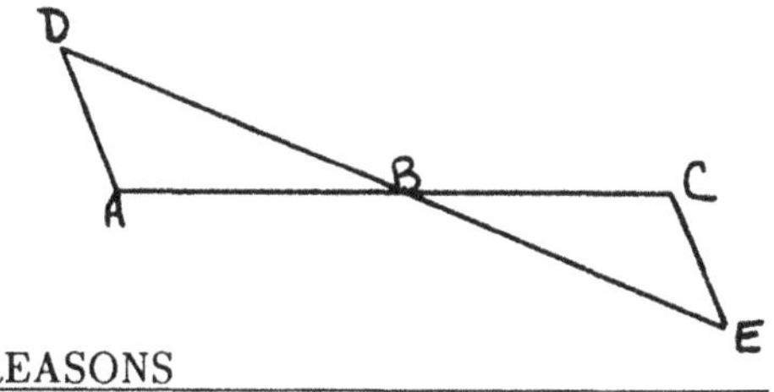

22. Given: $\overline{AC}$ and $\overline{DE}$ bisect each other at B.

Prove: $\overline{AD} \cong \overline{CE}$.

STATEMENTS	REASONS

CHAPTER 3

PARALLEL LINES

SUPPLEMENTARY EXERCISES

1. Given: $\overline{FA} \parallel \overline{DE}$ and $\overline{AF} \cong \overline{DE}$.
$\overline{AB} \cong \overline{DC}$.

Prove: $\angle F \cong \angle E$

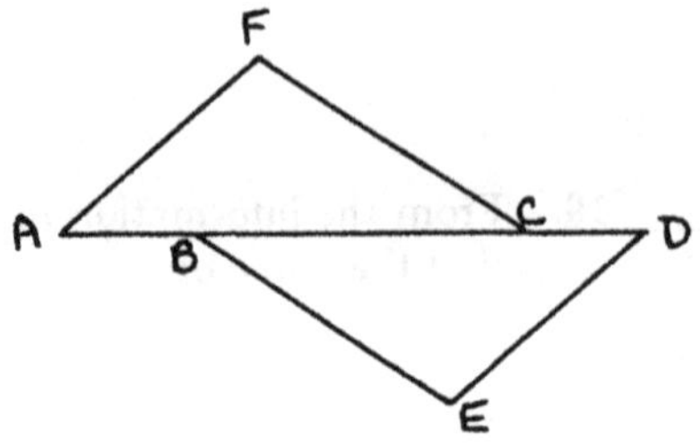

2. Given: $l_1 \parallel l_2$.

Prove: $\angle 1$ is supplementary to $\angle 2$.

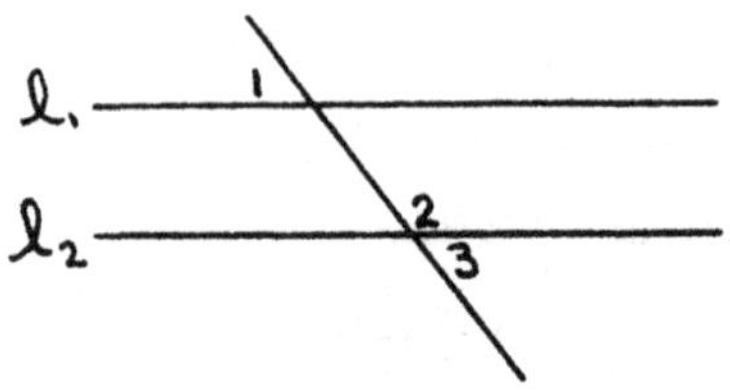

3. If $l_1 \parallel l_2$, find $m(\angle 1)$.

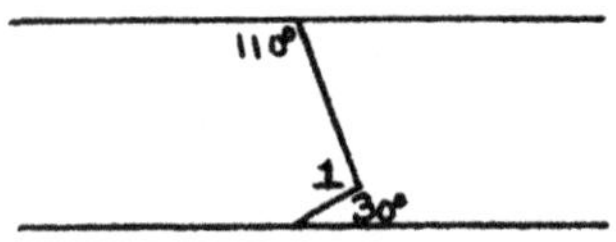

4. Given: $\overline{EA} \parallel \overline{DB}$ and $\overline{EA} \cong \overline{DB}$.
B is the midpoint of $\overline{AC}$.

Prove: $\overline{EB} \parallel \overline{DC}$.

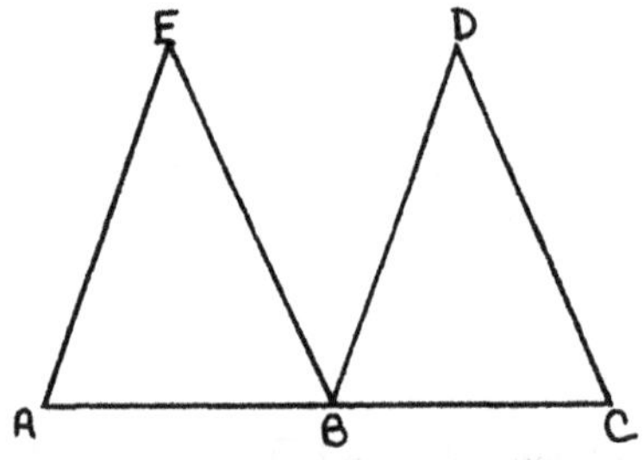

5. Given: $\angle 1 \cong \angle 2$.
$\overline{CD} \parallel \overline{AB}$.

Prove: $\triangle ABC$ is isosceles.

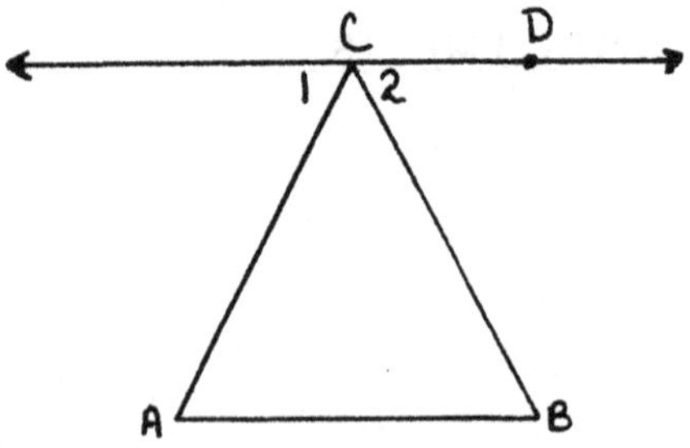

6. Given: $\overline{EA} \cong \overline{CB}$.

$\overline{EA} \perp \overline{AB}$, and $\overline{CB} \perp \overline{AB}$.

Prove: $\triangle ABD$ is isosceles.

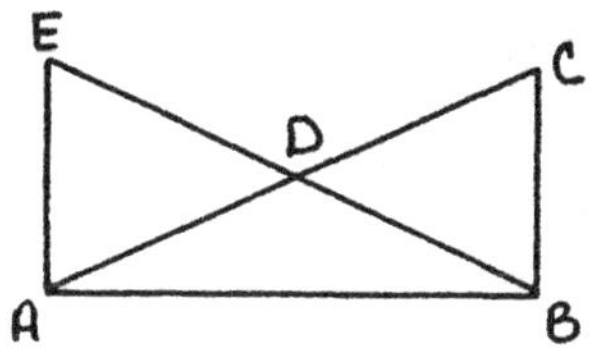

7. Given: $\overline{AB} \cong \overline{DC}$.

$\overline{EC} \cong \overline{FB}$.

$\angle 1 \cong \angle 2$.

Prove: $\overline{AE} \parallel \overline{DF}$.

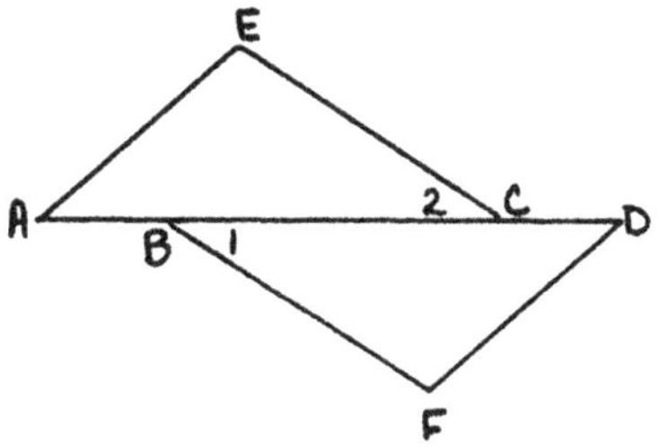

8. Prove: Segments joining the midpoints of the sides of a triangle divide the triangle into four congruent triangles.

9. If possible, construct a triangle given angles A and B and the nonincluded side $\overline{AC}$.

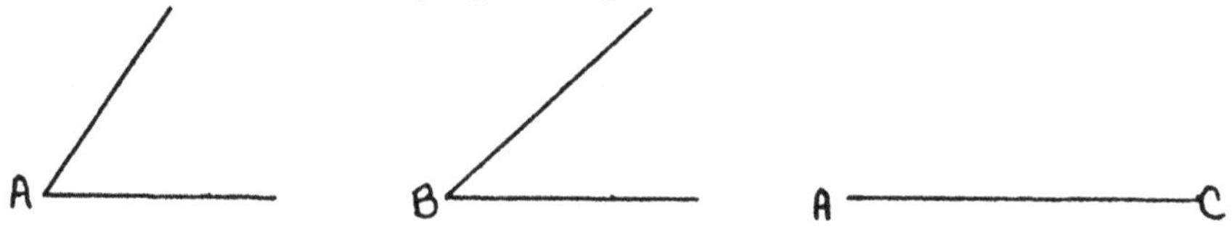

10. If possible, construct an isosceles triangle with base $\overline{AB}$ and vertex angle C.

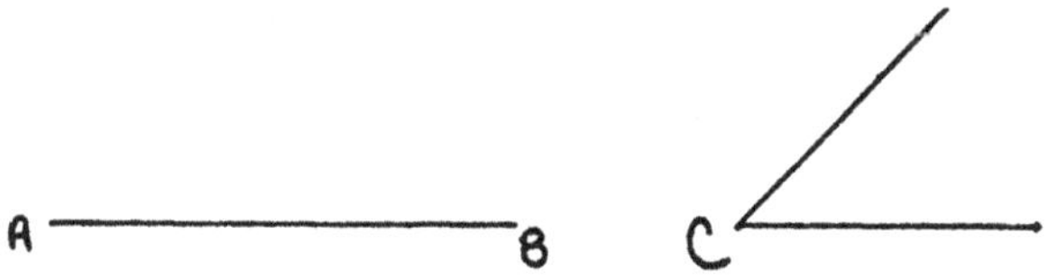

11. Use an indirect proof to prove there is no largest natural number.

LOGO SUPPLEMENT

This supplement is intended to give some basic commands of Logo and to demonstrate its usefulness working with graphics. The graphics are done with a small triangle, called the turtle, which you direct about the screen by giving it direction in angle measurements and a distance either forward or backward. Once a design has been written as a program and saved, it can be used in other designs. Some of the commands are:

COMMAND	ABBREVIATION	DESCRIPTION
TO (Program Name)		Starts the program editing. The prompt symbol "?" is now ">".
END		This is the last line of a program, denoting the end of the directions.
SAVE		Saves your program. Example: SAVE "Program
FORWARD *n*	FD *n*	Moves the turtle forward *n* steps. One step is approximately the thickness of a line.
BACK *n*	BK *n*	Moves the turtle back *n* steps.
RIGHT *n*	RT *n*	Turns the turtle *n* degrees to the right of the direction it is currently pointing.
LEFT *n*	LT *n*	Turns the turtle *n* degrees to the left of the direction it is currently pointing.
PENUP	PU	Picks the turtle up so its movement will not make marks on the screen.
PENDOWN	PD	Puts the turtle down so its movement will be seen.
REPEAT		Repeats a set of directions or program. Example: REPEAT 3[program name]
HIDETURTLE	HT	Makes the turtle invisible.
SHOWTURTLE	ST	Allows you to see the turtle. The turtle is initially in the middle of the screen pointing up. This is HOME.
EDIT		Allows you to edit programs to make additions or corrections. Press ESC to exit the edit mode. Example: EDIT "Program
CLEARTEXT	CT	Clears the text from the screen.
CLEARSCREEN	CS	Clears the text and the graphics.
FULLSCREEN		Allows the entire screen to be used by the turtle.
TEXTSCREEN		Allows the entire screen to be used for text.
SPLITSCREEN		Allows the turtle to be seen and 4 lines at the bottom for text.

EXAMPLE 1 Write a Logo program that will draw an equilateral triangle pointing up.

Solution It is helpful to imagine yourself "driving" the turtle as if it were a car and you were making the turns. It is very easy to get confused with the directions when the turtle is pointing down, so be sure to take your time and think it through. You may want to use graph paper to sketch out your design first, turning the paper so you and the turtle are facing the same direction as you go.

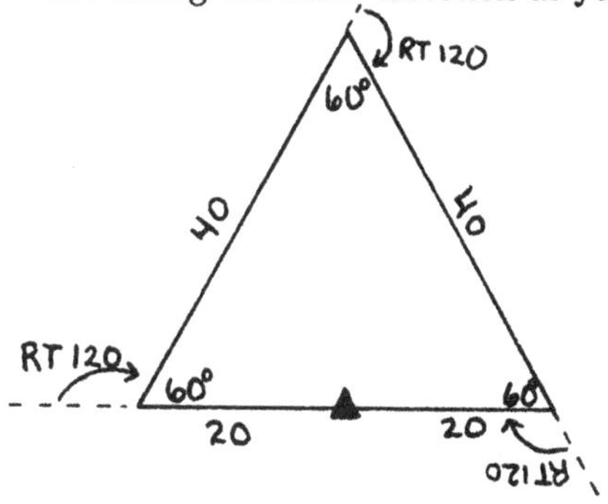


```
TO TRIANGLE
LT 90
FD 20
RT 120
FD 40
RT 120
FD 40
RT 120
FD 20
RT 90
END
```

EXERCISES

In exercises 1 – 4, follow the Logo commands to show what the turtle would draw on screen if the program were fed into a computer and run.

1.
```
TO EXER1
FD 50
RT 90
FD 50
RT 90
FD 50
RT 90
FD 50
RT 90
END
```

2.
```
TO EXER2
LT 45
FD 50
RT 90
FD 50
RT 90
FD 50
RT 90
FD 50
RT 135
END
```

3.
```
TO EXER3
FD 60
REPEAT 3[RT 90 FD 10]
RT 90
BK 50
END
```

4.
```
TO EXER4
REPEAT 6[EXER3 RT 60]
END
```

In exercises 5 – 8 write a Logo program to draw the following figures with the given lengths and angles. Make sure to end the program so that the turtle is placed and facing the direction shown.

5.

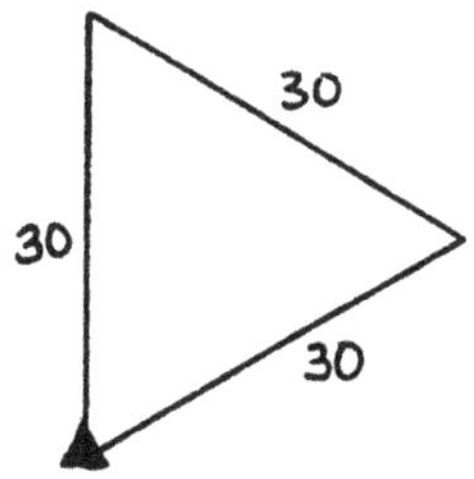

6.

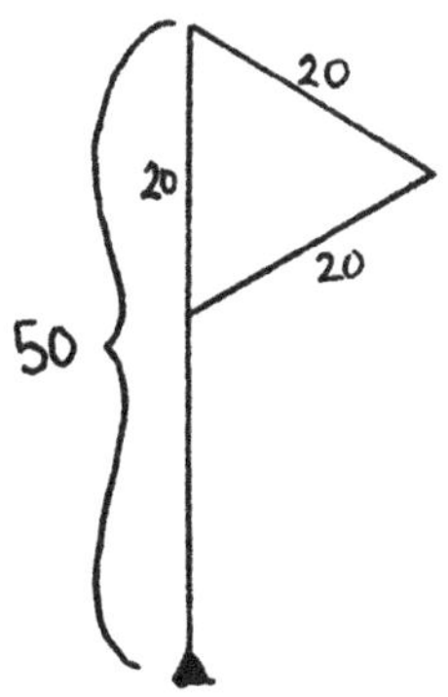

Chapter 3 Test

Form 1

Decide if each of the following statements is true or is false. Then mark T or F in the space provided.

_________ 1. If two distinct lines in the same plane are both perpendicular to a third line, the lines are parallel.

_________ 2. If two lines never intersect, then they are parallel.

_________ 3. Alternate interior angles are always equal.

_________ 4. If two lines are cut by a transversal, the sum of the measures of all interior angles is 360°.

_________ 5. If two lines are perpendicular, they are parallel to the same line.

_________ 6. If two angles in a triangle are 40° and 60°, the third angle must equal 80°.

_________ 7. If the base angles on an isosceles triangle are congruent to the base angles of another triangle, the vertex angles are congruent.

_________ 8. If the acute angles of one right triangle are congruent to the acute angles of another right triangle, the triangles are congruent.

_________ 9. The sum of the measures of the exterior angles of a triangle is equal to twice the sum of the measures of the interior angles of the triangle.

_________10. The two acute angles in a right triangle are supplementary.

_________11. The sum of the measures of the interior angles of a pentagon is 540°.

_________12. The sum of the measures of the exterior angles of a hexagon is 720°.

_________13. An interior angle of a regular octagon (an eight-sided figure with all sides and all angles congruent) has measure of 135°.

_________14. If the sum of the measures of the angles of a polygon is 1440°, the polygon is an octagon.

_________15. In an indirect proof, we always assume the opposite of what we are trying to prove.

Solve each problem and place the answer in the space provided.

_________16. $l \parallel r$ and $m \parallel n$. Find m($\angle 2$).

_________17. $l \parallel r$ and $m \parallel n$. Find m($\angle 1$).

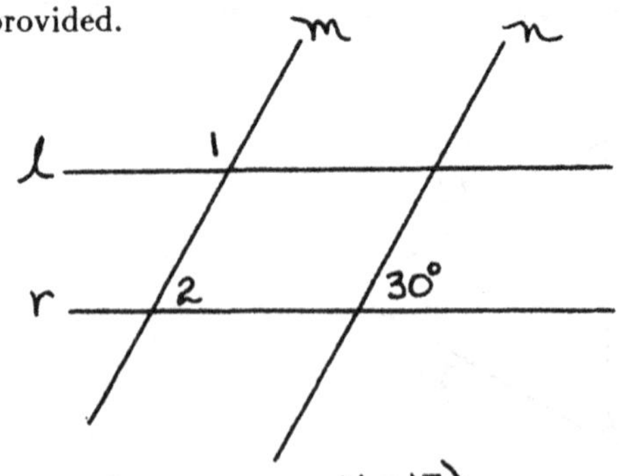

(Questions 16-17)

________18. Find m(∠CBD).

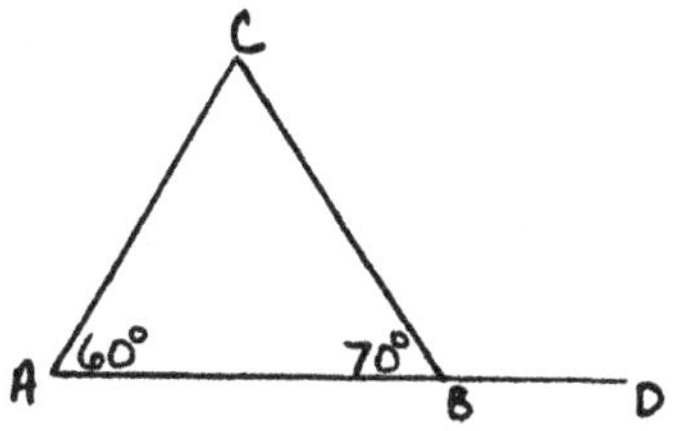

________19. Find the sum of the measures of the interior angles of a seven-sided polygon.

________20. Find the measure of an interior angle of a regular pentagon (five-sided figure).

________21. If one acute angle of a right triangle has measure of 56°, find the measure of the other acute angle.

________22. Are lines l and r parallel?

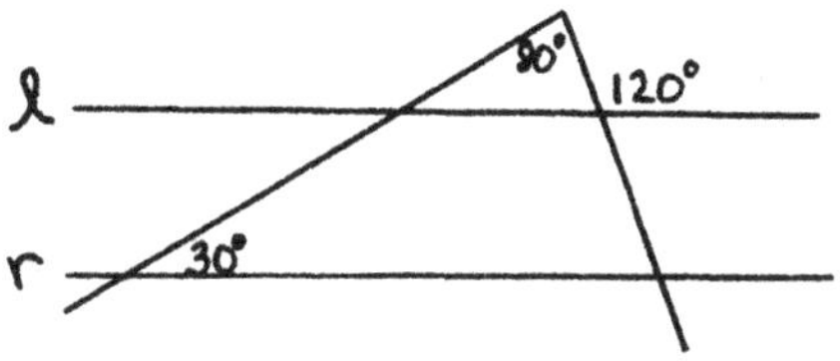

________23. If $\overline{AE} \parallel \overline{DC}$, find m(∠CBD).

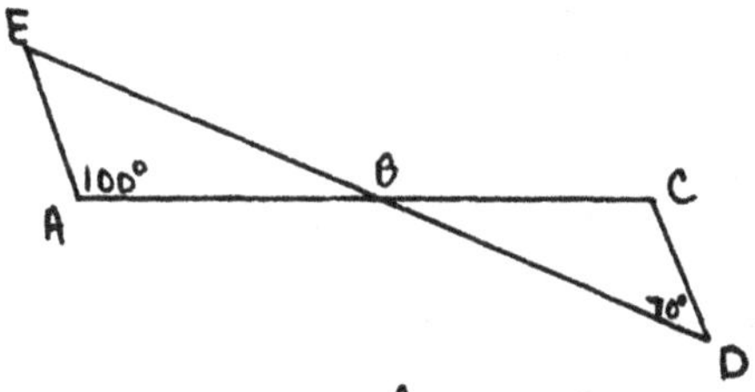

________24. $l_1 \parallel l_2$ and $l_3 \parallel l_4$. Find m(∠1).

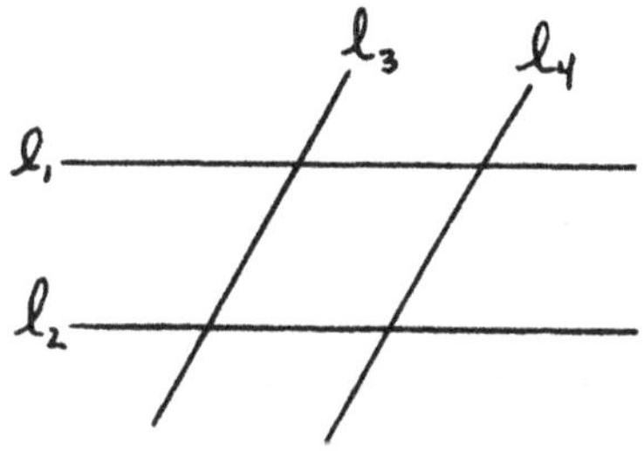

25. Given: △AFB ≅ △DCE.

$\overline{FD} \parallel \overline{AC}$.

Prove: $\overline{FB} \parallel \overline{EC}$.

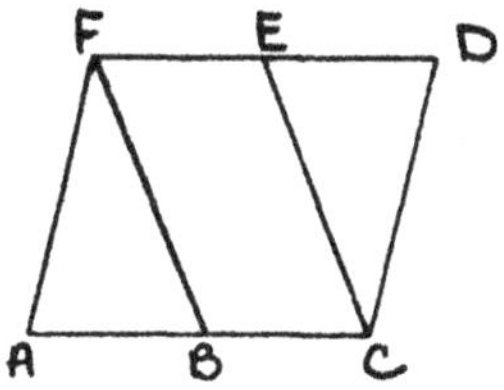

STATEMENTS	REASONS

Chapter 3 Test

Form 2

Decide if each of the following statements is true or is false. Then mark T or F in the space provided.

_________ 1. If two parallel lines are cut by a transversal, then interior angles on the same side of the transversal are complementary.

_________ 2. If two lines are cut by a transversal such that alternate interior angles are supplementary, the lines are parallel.

_________ 3. If a line is perpendicular to one of two perpendicular lines, then it is parallel to the other.

_________ 4. If two parallel lines are cut by a transversal, then interior angles on the same side of the transversal can never be complementary.

_________ 5. If one acute angle in a right triangle has measure of 45°, then the triangle is isosceles.

_________ 6. The measure of an exterior angle of a triangle is supplementary to the sum of the measures of the nonadjacent interior angles.

_________ 7. If any two angles and a side are congruent to two angles and a side of another triangle, the triangles are congruent.

_________ 8. If an acute angle and a leg of one right triangle are congruent to an acute angle and a leg of a second right triangle, then the triangles are congruent.

_________ 9. A triangle can have both a right angle and an obtuse angle.

_________ 10. If two sides and any angle in a triangle are congruent to two sides and any angle in a second triangle, the triangles are congruent.

_________ 11. The sum of the measures of the interior angles of a hexagon (a six-sided polygon) is the same as the sum of the measures of the exterior angles of a triangle.

_________ 12. The sum of the measures of the interior angles of a pentagon(a 5-sided polygon) is 540°.

_________ 13. An interior angle of a regular pentagon (a five-sided polygon with all sides and all angles congruent) has measure of 18°.

_________ 14. If the measure of one interior angle of a regular polygon is 90°, the polygon is a square.

_________ 15. In an indirect proof, we always state the given information first.

Solve each problem and place the answer in the space provided.

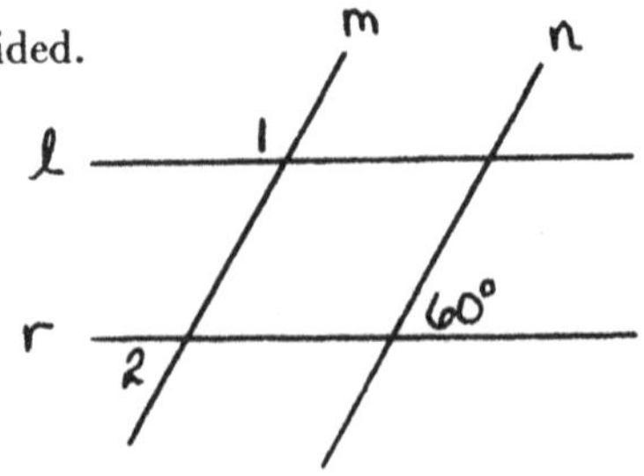

_________ 16. $l \parallel r$ and $m \parallel n$. Find m(∠2).

_________ 17. $l \parallel r$ and $m \parallel n$. Find m(∠1).

________18. In $\triangle ABC$, $\overline{AC} \cong \overline{BC}$. Find m($\angle$CBD).

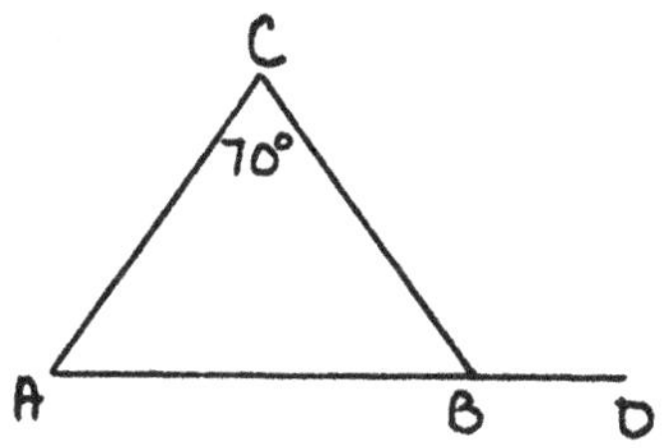

________19. Find the sum of the measures of the interior angles of a nine-sided polygon.

________20. Find the measure of an interior angle of a regular seven-sided polygon.

________21. $\overline{AB} \cong \overline{CB}$ and $\overline{DB} \cong \overline{EB}$. Are $\overline{AE}$ and $\overline{CD}$ parallel?

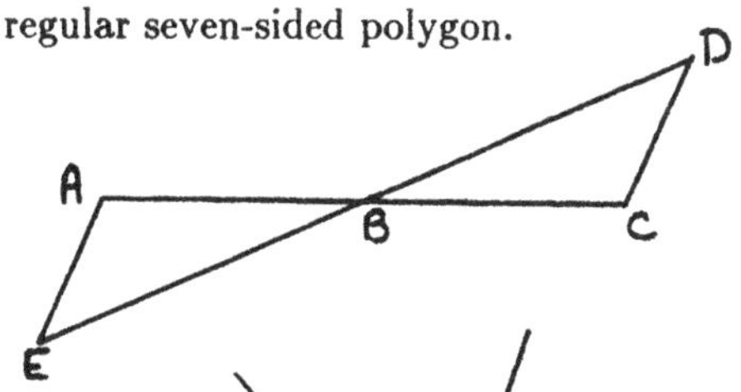

________22. If $l_1 \parallel l_2$, find m($\angle$1).

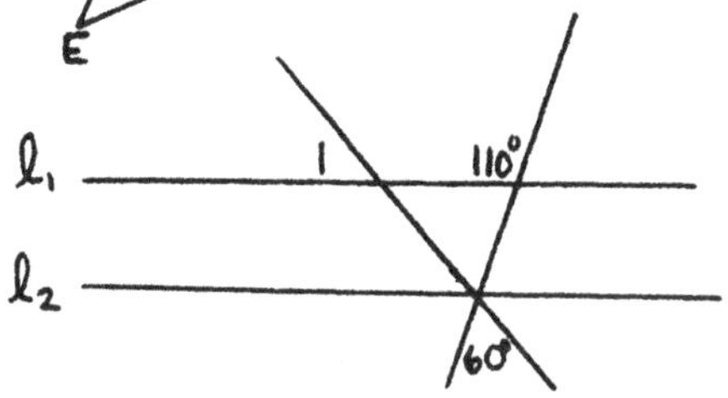

________23. $\triangle ABC$ is a right triangle and $\overline{BD} \parallel \overline{AC}$. Find m($\angle$C).

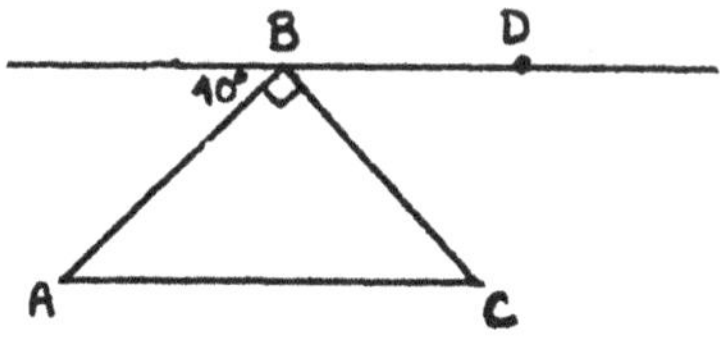

(Questions 23-24)

________24. $\triangle ABC$ is a right triangle and $\overline{BD} \parallel \overline{AC}$. Find m($\angle$ABD).

25. Given: $\angle 1 \cong \angle 2$.
$\overline{CD} \parallel \overline{AB}$.

Prove: $\triangle ABC$ is isosceles.

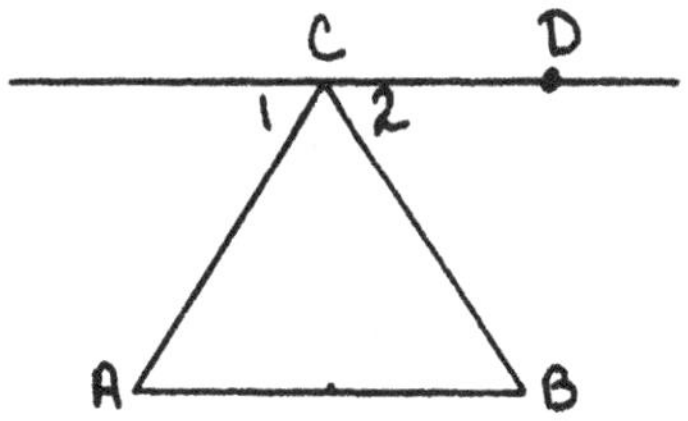

STATEMENTS	REASONS

Chapter 3 Test

Form 3

Decide if each of the following statements is true or is false. Then mark T or F in the space provided.

_________ 1. If two lines are cut by a transversal, corresponding angles congruent.

_________ 2. If two lines are cut by a transversal such that alternate exterior angles are congruent, the lines are parallel.

_________ 3. If two lines are cut by a transversal such that interior angles on the same side of the line are supplementary, the lines are parallel.

_________ 4. If two parallel lines are cut by a transversal such that the transversal is perpendicular to one of the lines, it is parallel to the other.

_________ 5. The acute angles of a right triangle are supplementary.

_________ 6. An exterior angle of a triangle is complimentary to the sum of the measures of the non-adjacent interior angles.

_________ 7. An exterior angle of a triangle is the same as the sum of the measures of the nonadjacent interior angles.

_________ 8. Through a point not on the line there exists infinitely many parallel lines to the line.

_________ 9. If a triangle has an obtuse angle, the other two angles must be acute angles.

_________10. If two isosceles triangles are congruent, then their vertex angles are congruent.

_________11. The sum of the measures of the interior angles of a septagon (a 7-sided polygon) is 1260°.

_________12. The measure an interior angles of a regular quadrilateral is 90°.

_________13. If the sum of the measures of the interior angles of a polygon is 1980°, the polygon has 13 sides.

_________14. The sum of the measures of the exterior angles of a hexagon is 360°.

_________15. In an indirect proof you make an assumption that is the opposite of what you are trying to prove and look for a contradiction.

Solve each problem and place the answer in the space provided.

_________16. $l \parallel r$ and $m \parallel n$. Find m($\angle 1$).

_________17. $l \parallel r$ and $m \parallel n$. Find m($\angle 2$).

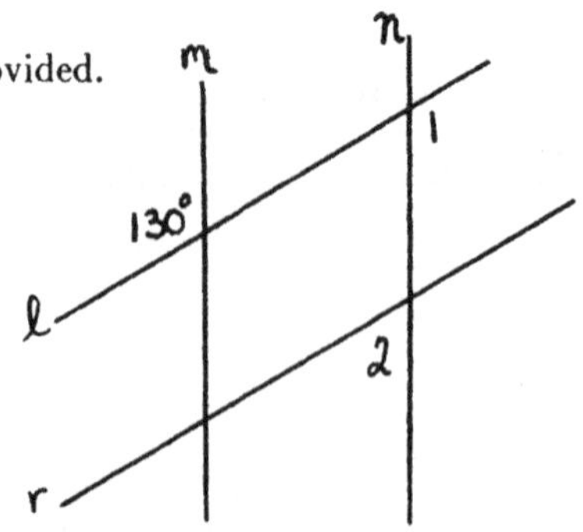

(Questions 16-17)

_________18. In $\triangle ABC$, $\overline{AC} \cong \overline{BC}$. Find m($\angle$CBD).

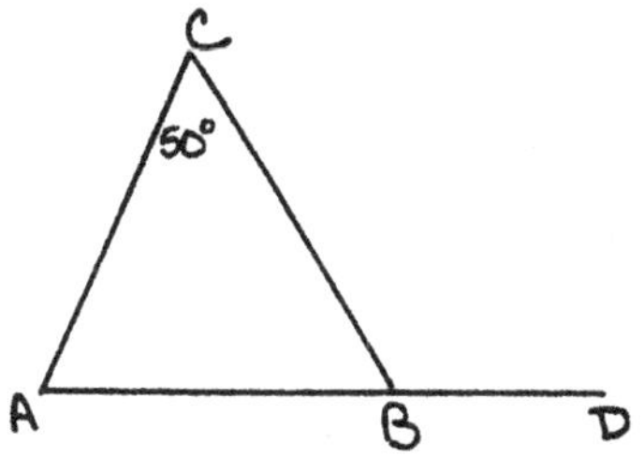

_________19. Find the sum of the measures of the interior angles of an octagon (an eight-sided polygon).

_________20. Find the measure of an interior angle of a regular hexagon (a six-sided polygon).

_________21. $\overline{AE} \cong \overline{CD}$ and $\overline{AB} \cong \overline{CB}$. Are $\overline{AE}$ and $\overline{CD}$ parallel?

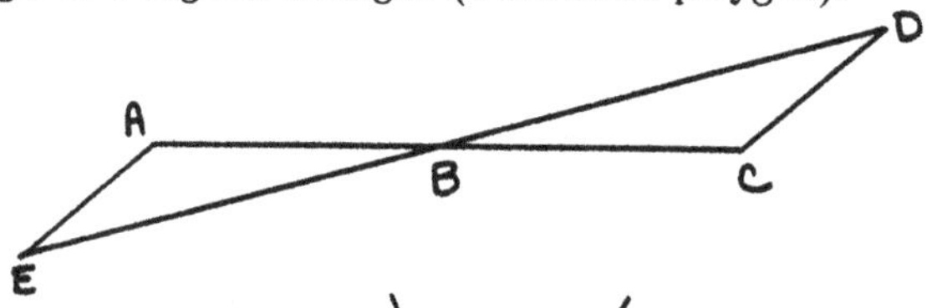

_________22. If $l_1 \parallel l_2$, find m($\angle$1).

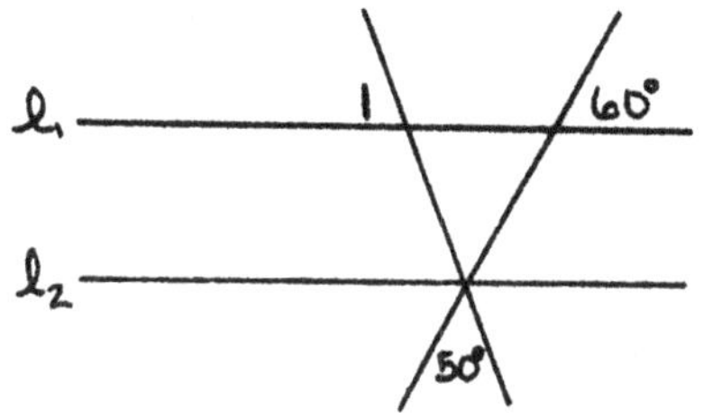

_________23. $\triangle ABC$ is a right triangle and $\overline{BD} \parallel \overline{AC}$. Find m($\angle$A).

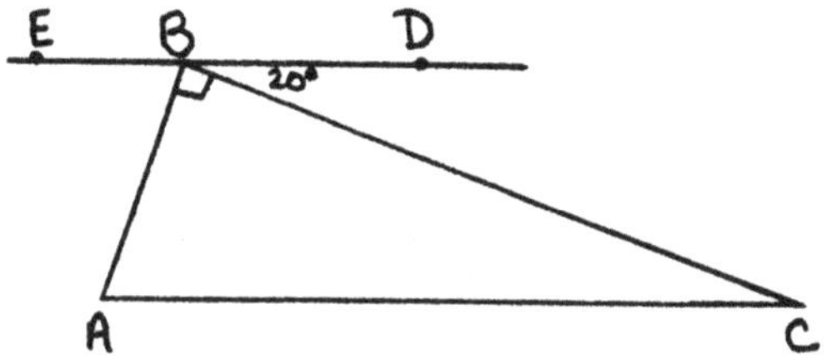

(Questions 23-24)

_________24. $\triangle ABC$ is a right triangle and $\overline{BD} \parallel \overline{AC}$. Find m($\angle$ABE).

25. Given: $\overline{AC} \cong \overline{BC}$.
$\overline{CD} \parallel \overline{AB}$.

Prove: $\angle 1 \cong \angle 2$.

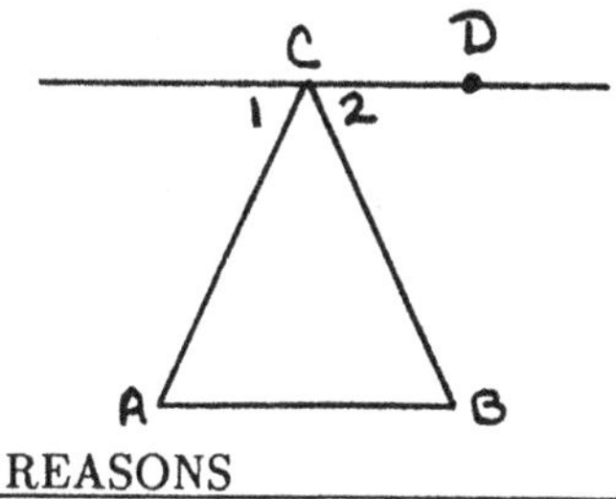

STATEMENTS	REASONS

CHAPTER 4

PARALLELOGRAMS AND TRAPEZOIDS

SUPPLEMENTARY EXERCISES

1. Given: ▱ABCD, with $\overline{DM}$ and $\overline{BN}$ perpendicular to diagonal $\overline{AC}$.

   Prove: $\overline{DM} \cong \overline{BN}$.

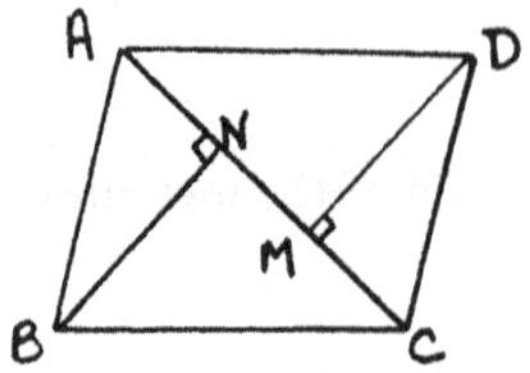

2. Prove: The line segments joining the midpoints of the sides of a square form another square.

3. Prove: If two consecutive angles of a quadrilateral are right angles, then the bisectors of the other angles are perpendicular.

4. Prove: Segments joining the midpoints of the diagonals of a trapezoid in one-half the difference of the bases. (Hint: If extended, this segment bisects the nonparallel sides of the trapezoid.)

5. Given: ▱ABCD.
   $\angle 1 \cong \angle 2$.

   Prove: $\overline{EC} \cong \overline{FA}$.

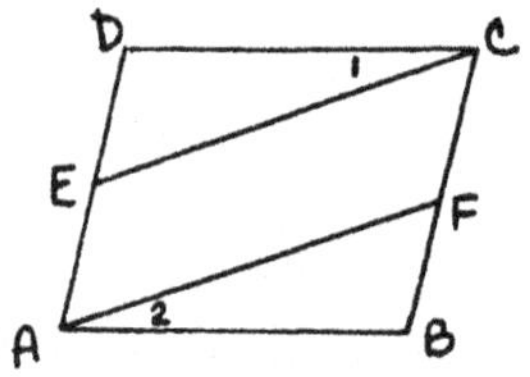

6. Given: ABCD is a rectangle.
   $\overline{AE} \cong \overline{BE}$.

   Prove: $\triangle$ CED is isosceles.

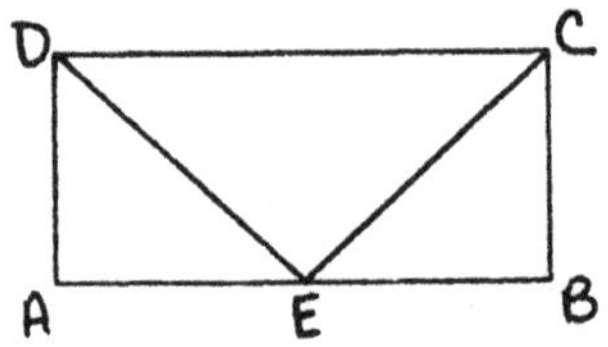

7. Given: ▱ABCD
   $\overline{DE} \perp \overline{AF}$.
   $\overline{CF}$ is an altitude of $\triangle$ CFB.

   Prove: DEFC is a rectangle.

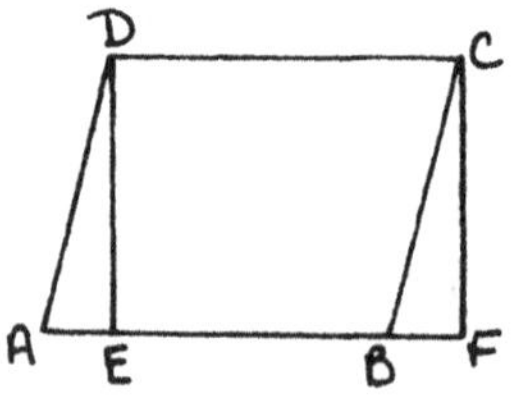

8. Given: ▱ABCD.
   $\angle GHA \cong \angle FEC$.
   $\overline{HB} \cong \overline{ED}$.

   Prove: $\overline{AG} \cong \overline{CF}$.

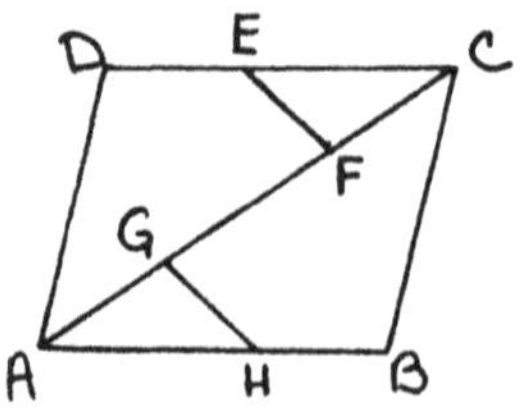

9. If possible, construct a rhombus given side $\overline{AB}$ and $\angle A$.

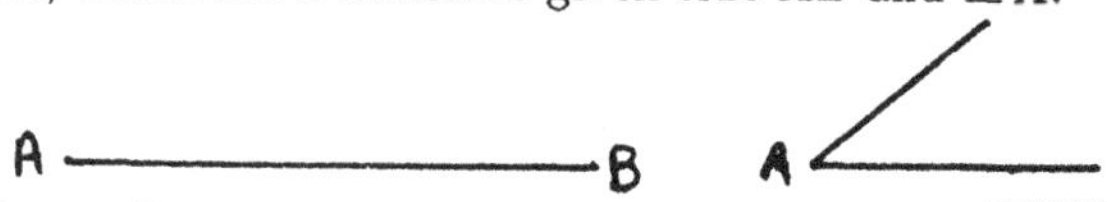

10. $\overline{MN}$ is the median of an isosceles trapezoid with side $\overline{AMB}$ and base angle A. Construct the trapezoid.

LOGO EXERCISES FOR CHAPTER FOUR

In exercises 1 – 4, follow the Logo commands to show what the turtle would draw on the screen if the program were fed into a computer and run. State whether the figure is a parallelogram, rhombus, rectangle, square, or a trapezoid. Some figures may fit into more than one category.

1.	2.	3.	4.
TO EXER1	TO EXER2	TO EXER3	TO EXER4
RT 30	RT 50	RT 20	LT 30
FD 30	FD 30	FD 40	FD 18
RT 60	RT 40	RT 90	LT 60
FD 50	FD 30	FD 40	FD 30
RT 120	RT 140	RT 90	LT 60
FD 30	FD 30	FD 40	FD 18
RT 60	RT 40	RT 90	LT 120
FD 50	FD 30	FD 40	FD 48
RT 90	RT 90	RT 70	LT 90
END	END	END	END

In exercises 5 – 8 write a Logo program to draw the following figures with the given lengths and angles.

5.

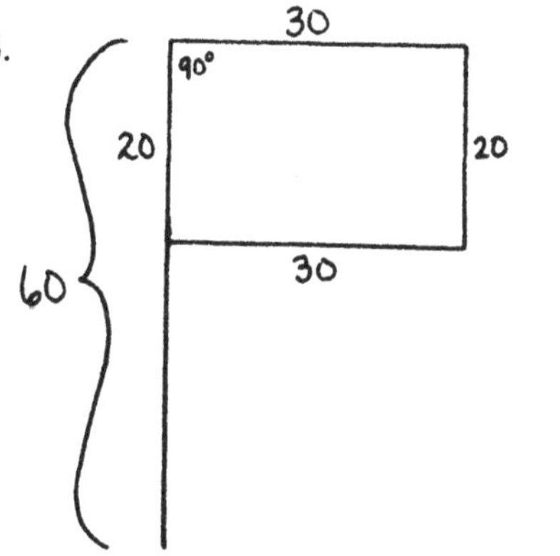

6.

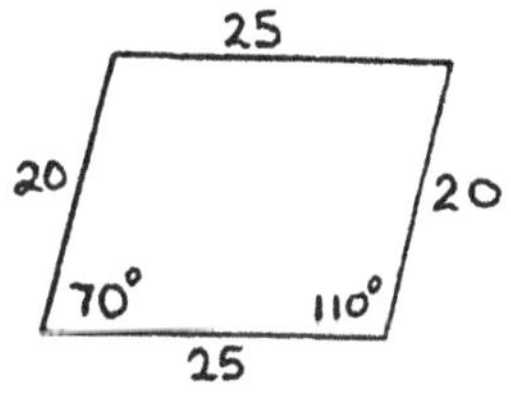

7.

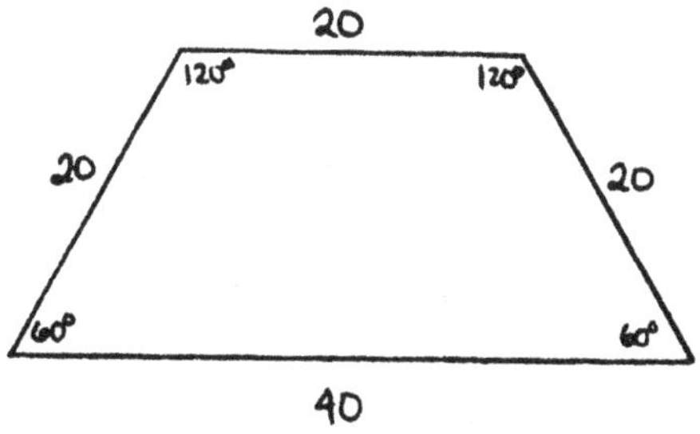

8.

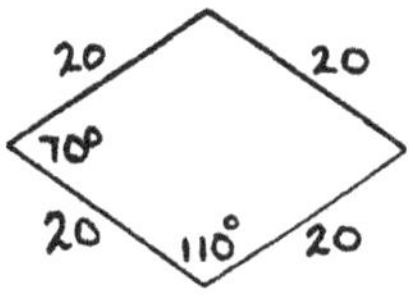

Chapter 4 Test

Form 1

Decide if each of the following statements is true or is false. Then mark T or F in the space provided.

_________ 1. A diagonal of a rhombus divides the rhombus into two congruent triangles.

_________ 2. Consecutive angles of a quadrilateral are supplementary.

_________ 3. A quadrilateral is a square.

_________ 4. A parallelogram has exactly one pair of parallel sides.

_________ 5. The diagonals of a parallelogram are congruent.

_________ 6. The diagonal of a parallelogram divides the parallogram into two congruent triangles.

_________ 7. Opposite angles of a parallelogram are supplementary.

_________ 8. The diagonals of a rhombus bisect each other.

_________ 9. If a line joins the midpoints of two sides of a triangle, then that line is equal to one-half the measure of the third side.

_________10. A rhombus is a square.

_________11. A square is a rectangle.

_________12. The diagonals of a rectangle are congruent.

_________13. The diagonals of a rectangle are perpendicular.

_________14. A trapezoid has two nonparallel sides.

_________15. The median of the trapezoid is parallel to its bases.

Solve each problem and place the answer in the space provided.

_________16. $\triangle ABC$ is isosceles with $\overline{AC} \cong \overline{BC}$ and M and N are midpoints of $\overline{AC}$ and $\overline{BC}$ respectively. Find $m(\angle CMN)$.

_________17. $\triangle ABC$ is isosceles with $\overline{AC} \cong \overline{BC}$ and M and N are midpoints of $\overline{AC}$ and $\overline{BC}$ respectively. Find $m(\overline{MN})$.

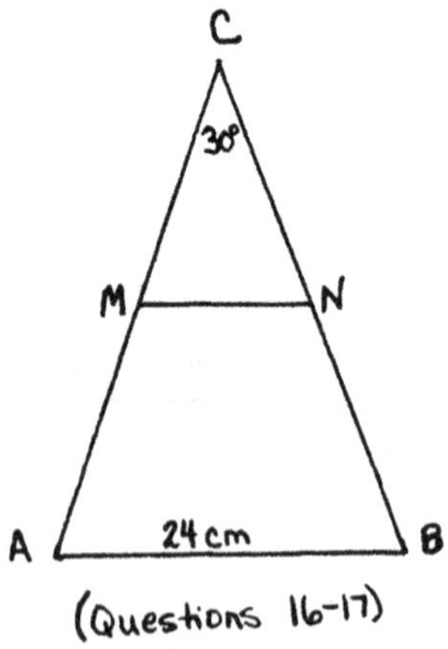

_________18. ABCD is a trapezoid. Find the measure of $\overline{MN}$, the median of ABCD.

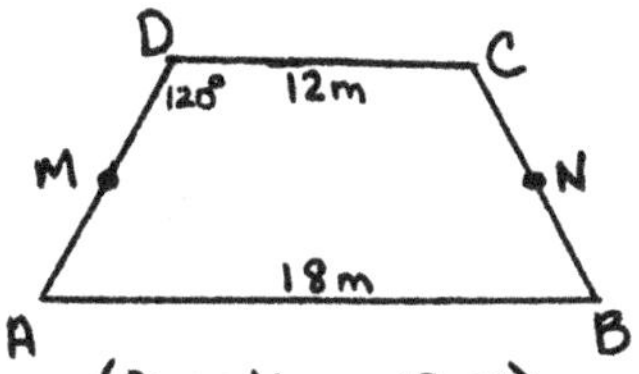

(Questions 18-19)

_________19. ABCD is an isosceles trapezoid with $m(\angle D) = 120°$. Find $m(\angle B)$.

_________20. ABCD is a parallelogram with $m(\angle A) = x°$ and $m(\angle D) = (3x - 4)°$. Find $m(\angle C)$.

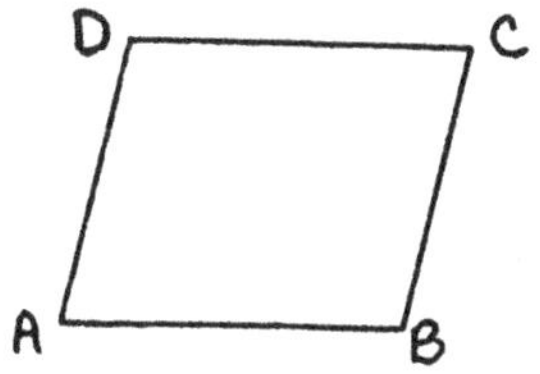

_________21. ABCD is a rectangle, and

$m(\overline{DA}) = 2x - 4$ $\quad$ $m(\overline{BC}) = \frac{1}{2}x + 8$

$m(\overline{AB}) = 3y + 4$ $\quad$ $m(\overline{CD}) = y + 12.$

Find the perimeter of $\square$ABCD.

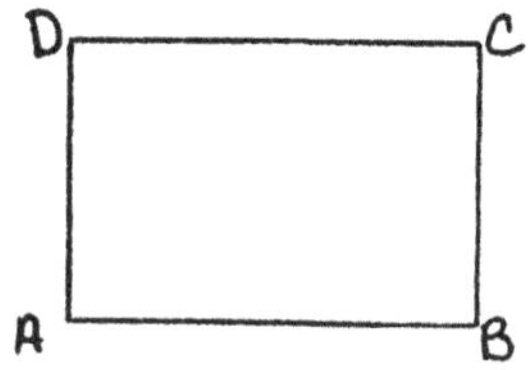

_________22. ABCD is a square, and

$m(\overline{AC}) = 3x + 5$

$m(\overline{BD}) = 2x + 6.$

Find x.

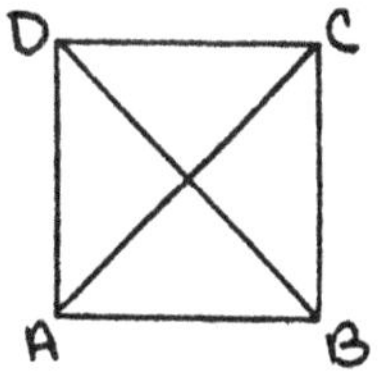

_________23. ABCD is a quadrilateral with E, F, G, and H midpoints of sides $\overline{DA}$, $\overline{AB}$, $\overline{BC}$, and $\overline{CD}$ respectively. If $m(\angle E) = 30°$, find $m(\angle H)$.

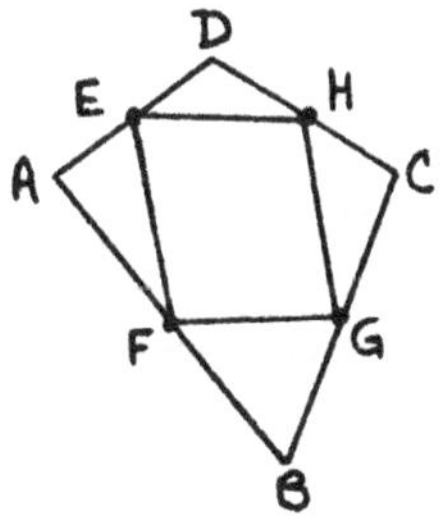

24. Given: Quadrilateral ABCD with $\angle B \cong \angle D$ and $\angle A \cong \angle C$.

Prove: ABCD is a parallelogram.

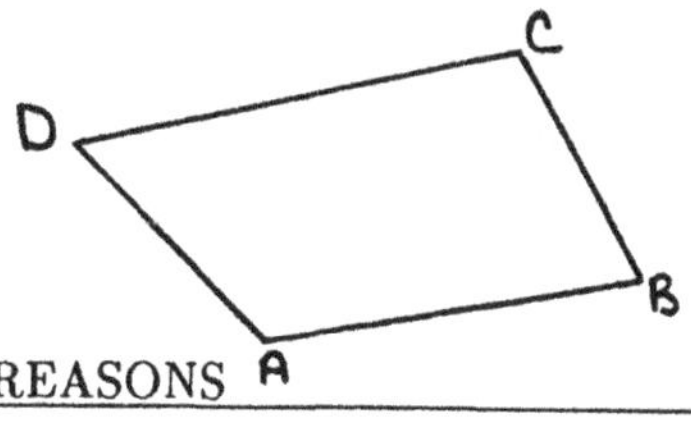

STATEMENTS	REASONS

Chapter 4 Test

Form 2

Decide if each of the following statements is true or is false. Then mark T or F in the space provided.

________ 1. The diagonals of a quadrilateral bisect each other.

________ 2. Opposite angles of a quadrilateral are congruent.

________ 3. If one pair of opposite sides of a quadrilateral are parallel and the other pair of opposite sides are congruent, the quadrilateral is a parallelogram.

________ 4. If the diagonals of a parallelogram are congruent, the parallelogram is a rectangle.

________ 5. If the diagonals of a quadrilateral bisect each other, the quadrilateral is a parallelogram.

________ 6. A rectangle is a rhombus.

________ 7. Consecutive angles of a rhombus are supplementary.

________ 8. The diagonals of a rhombus bisect each other.

________ 9. A square is a rhombus.

________10. The bisectors of the angles of a rhombus are the diagonals the rhombus.

________11. If a line joins the midpoints of the two sides of a triangle, the smaller triangle formed is isosceles.

________12. A parallelogram is a trapezoid.

________13. The measure of the median of a trapezoid is one-half the sum of the measures of the bases.

________14. If the base angles of a trapezoid are congruent, the trapezoid is isosceles.

________15. A square cannot be a trapezoid.

Solve each problem and place the answer in the space provided.

________16. $\triangle ABC$ is isosceles with $\overline{AC} \cong \overline{BC}$, and M and N are midpoints of $\overline{AC}$ and $\overline{BC}$ respectively. Find $m(\angle C)$.

________17. $\triangle ABC$ is isosceles with $\overline{AC} \cong \overline{BC}$ and M and N are midpoints of $\overline{AC}$ and $\overline{BC}$ respectively. Find $m(\overline{MN})$.

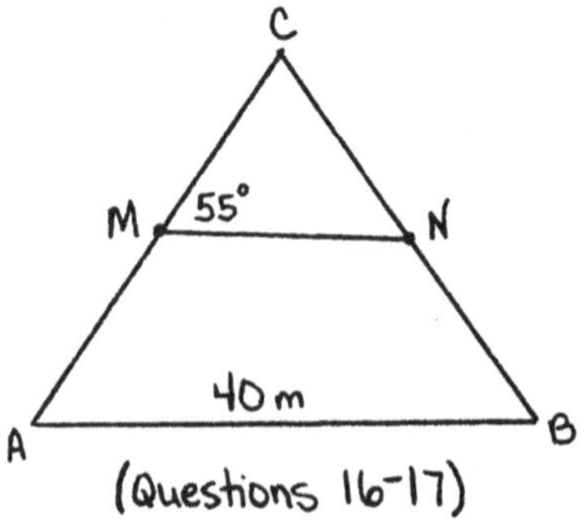

(Questions 16-17)

________18. ABCD is a trapezoid with E and F the midpoints of $\overline{AD}$ and $\overline{BC}$ respectively. Find $m(\overline{AB})$.

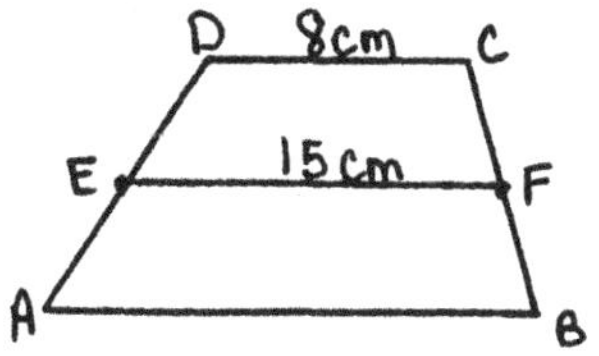

________19. ABCD is an isosceles trapezoid with $m(\angle A) = 40°$. Find $m(\angle C)$.

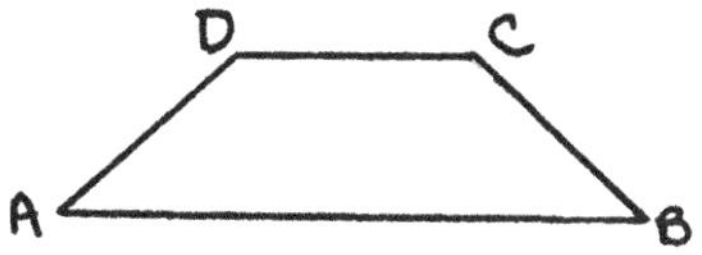

________20. ABCD is a rhombus with diagonal $\overline{BD}$. If $m(\angle A) = 110°$, find $m(\angle CDB)$.

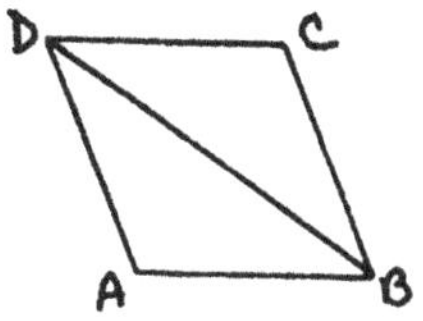

________21. ABCD is a parallelogram with

$m(\angle B) = (3x + 6)°$

$m(\angle D) = (2x + 21)°$

Find $m(\angle A)$.

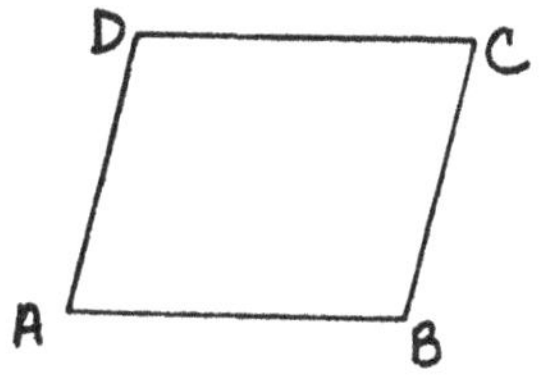

________22. ABCD is a parallelogram, and

$m(\overline{DA}) = 4x - 8$ $m(\overline{BC}) = x + 16$

$m(\overline{AB}) = 3y + 4$ $m(\overline{CD}) = y + 12$.

Find the perimeter of parallelogram ABCD.

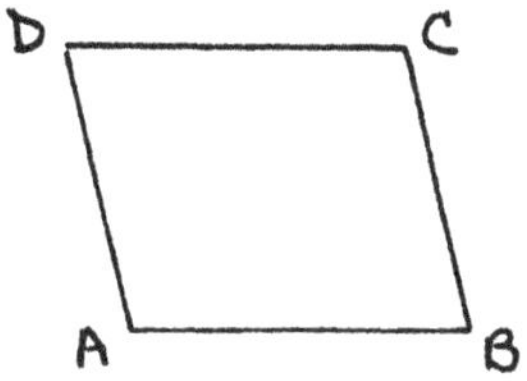

________23. Given ABCD is a rectangle with E, F, G, and H midpoints of sides $\overline{DA}$, $\overline{AB}$, $\overline{BC}$, and $\overline{CD}$ respectively. If $m(\angle HEF) = 30°$, find $m(\angle GHE)$.

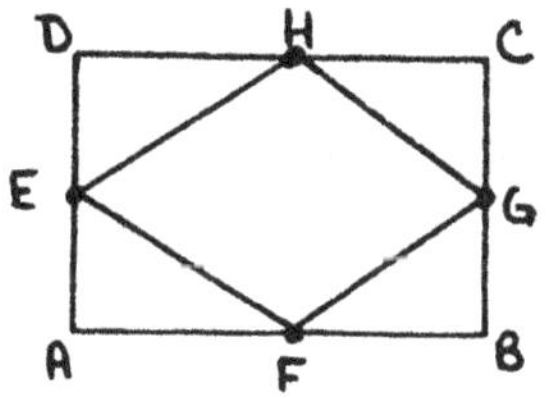

24. Given: ABDE is a parallelogram and $\overline{BC}$ is the base of isosceles triangle BCD.

Prove: ACDE is an isosceles trapezoid.

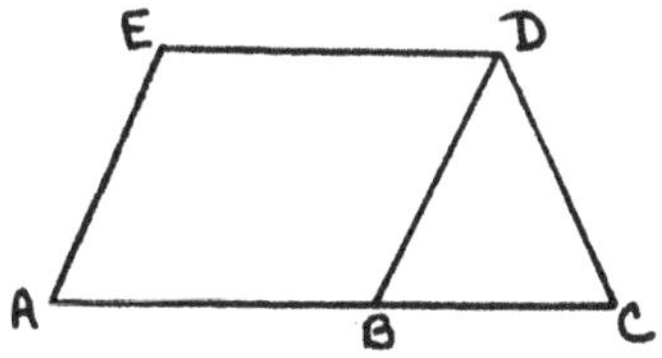

STATEMENTS	REASONS

Chapter 4 Test

Form 3

Decide if each of the following statements is true or is false. Then mark T or F in the space provided.

_________ 1. A parallelogram is a quadrilateral.

_________ 2. A quadrilateral has congruent opposite angles.

_________ 3. If both pairs of opposite sides are congruent, then the quadrilateral is a parallelogram.

_________ 4. The distance between two parallel lines depends on the slant of the transversal.

_________ 5. If both pairs of opposite sides of a quadrilateral are parallel, the quadrilateral is a rhombus.

_________ 6. The diagonals of a parallelogram are perpendicular.

_________ 7. A rectangle is a rhombus.

_________ 8. One pair of opposite sides of a rectangle are perpendicular to the other pair of opposite sides.

_________ 9. The diagonals of a rhombus are perpendicular.

_________10. All squares are rectangles.

_________11. The diagonals of a rectangle are perpendicular.

_________12. A diagonal of a square divides the square into two congruent isosceles triangles.

_________13. Nonconsective angles of an isosceles trapezoid are supplementary.

_________14. It is possible to trisect a line segment using a compass and a straightedge.

_________15. The measure of the median of a trapezoid is equal to one-half the measure of the base.

Solve each problem and place the answer in the space provided.

_________16. $\triangle ABC$ is isosceles with $\overline{AC} \cong \overline{BC}$, and M and N are midpoints of $\overline{AC}$ and $\overline{BC}$ respectively. Find $m(\angle M)$.

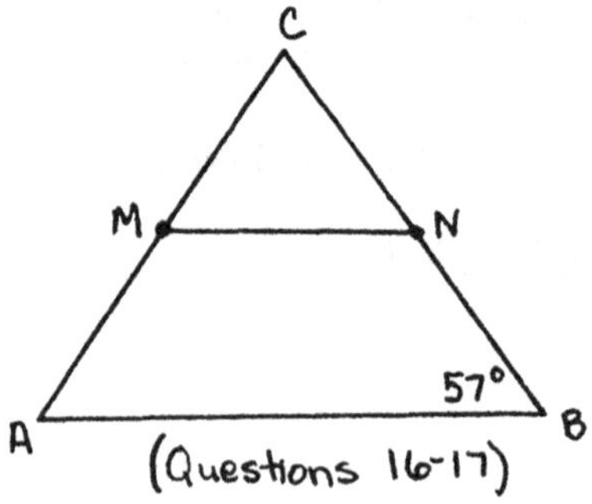

_________17. $\triangle ABC$ is isosceles with $\overline{AC} \cong \overline{BC}$ and M and N are midpoints of $\overline{AC}$ and $\overline{BC}$ respectively. If the perimeter of $\triangle ABC$ is 62 cm and $m(\overline{AC}) = 25$ cm, Find $m(\overline{MN})$.

_________18. ABCD is a trapezoid with E and F the midpoints of $\overline{AD}$ and $\overline{BC}$ respectively. Find m($\overline{DC}$).

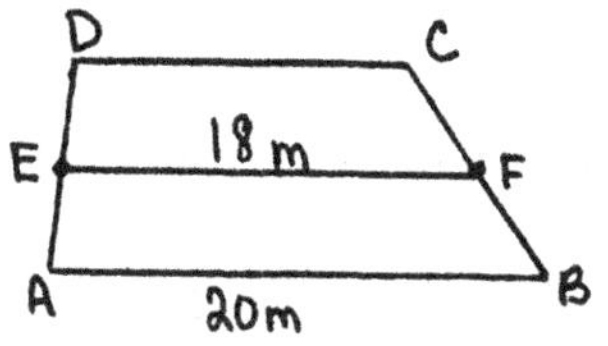

_________19. ABCD is an isosceles trapezoid with
m($\angle$A) = $(3x+6)^\circ$
m($\angle$B) = $(2x+21)^\circ$
Find m($\angle$A).

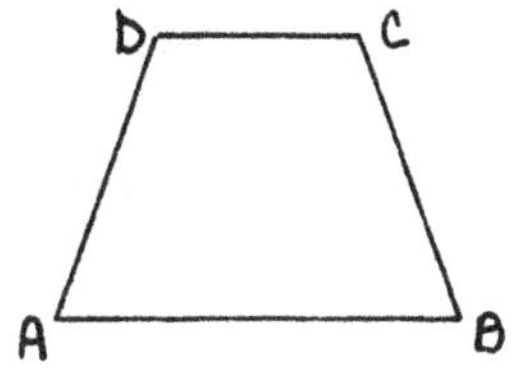

_________20. $l_1 \parallel l_2 \parallel l_3$. Find the distance between l_1 and l_3.

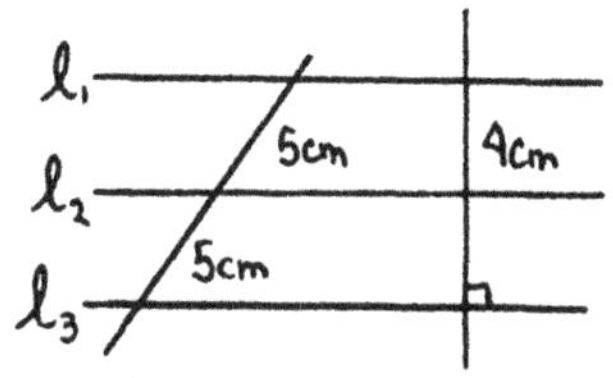

_________21. ABCD is a rhombus with diagonal $\overline{BD}$. If m($\angle$BDC) = 30°, find m($\angle$A).

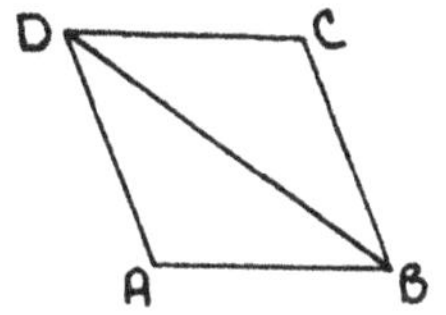

_________22. ABCD is a parallelogram, and

m($\overline{DA}$) = $4x-8$ m($\overline{BC}$) = $x+16$

m($\overline{AB}$) = $6y+8$ m($\overline{CD}$) = $2y+24$.

Find the perimeter of parallelogram ABCD.

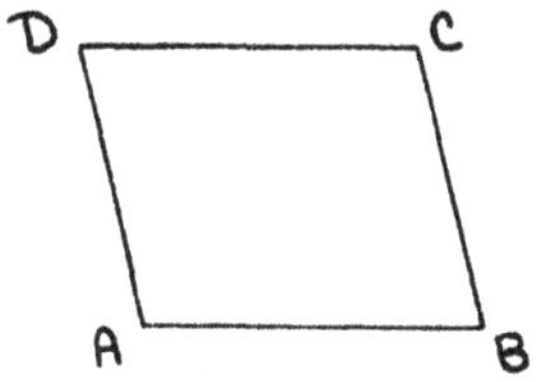

_________23. Given ABCD is a rectangle with E, F, G, and H midpoints of sides $\overline{DA}$, $\overline{AB}$, $\overline{BC}$, and $\overline{CD}$ respectively. If m($\angle$DEH) = 60°, find m($\angle$FGH).

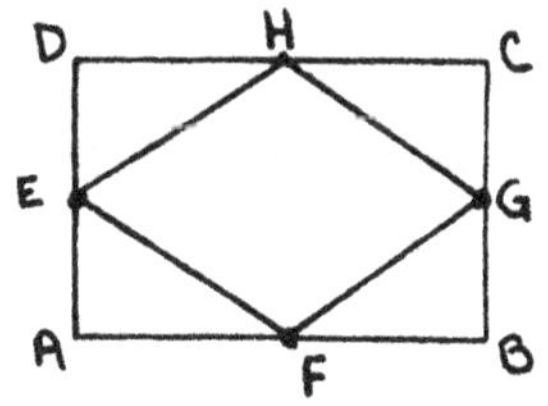

24. Given: $\triangle$ABC is an isosceles triangle with M and N midpoints of $\overline{AC}$ and $\overline{BC}$ respectively.

Prove: ABNM is an isosceles trapezoid.

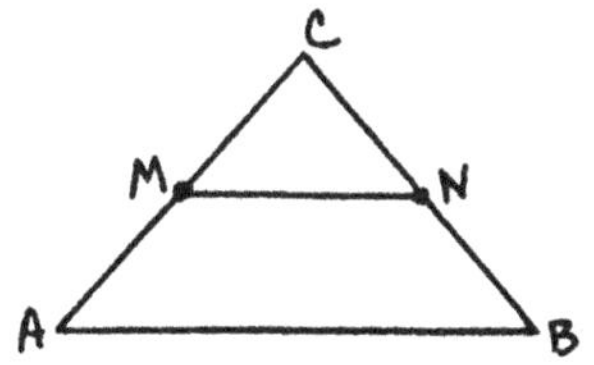

STATEMENTS	REASONS

CHAPTER 5

AREAS AND RELATED TOPICS

SUPPLEMENTARY EXERCISES

1. Given: $\triangle ABC$, with $\overline{DG}$ intersecting $\overline{AB}$ and $\overline{CB}$ at midpoints E and F. DACG is a rectangle. $\overline{BI} \perp \overline{AC}$.

   Prove: A(DACG) = A($\triangle$ ABC).

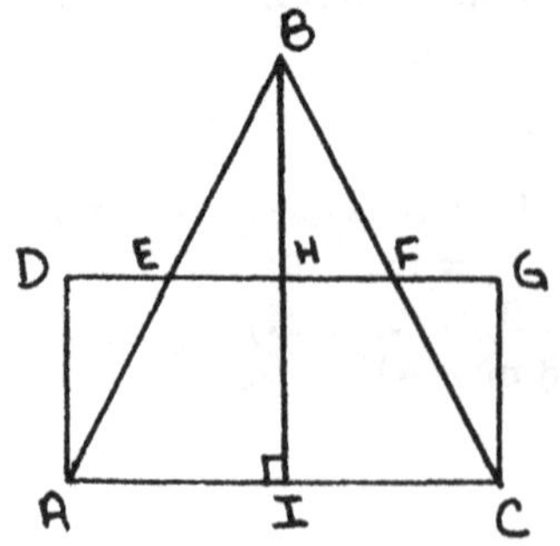

2. If the side of a square is doubled, what change occurs in the area? (multiplied by 4)

3. A rectangle has a side of length 12 and a diagonal of length 13. Find its area. (60 sq units)

4. The distance around a rhombus, called its *perimeter*, is 24 cm. Find the area of the rhombus if one of its angles is 30°. (18 sq cm)

5. Prove that the median of a triangle divides the triangle into two triangles of equal area.

6. Given: $\triangle$ ABC and $\triangle$ ABD have equal areas.

   Prove: $\overline{CD}$ is bisected by $\overline{AB}$.

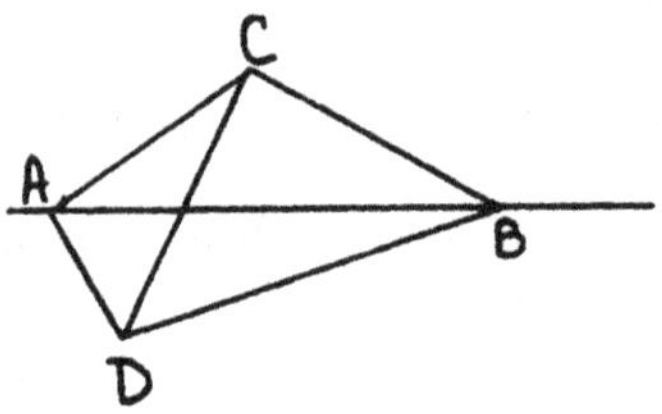

7. Prove: The diagonals of a parallelogram divide it into four nonoverlapping triangles of equal area.

8. Given: Squares ABCD and ACEF.

   Prove: A(ACEF) = 2A(ABCD).

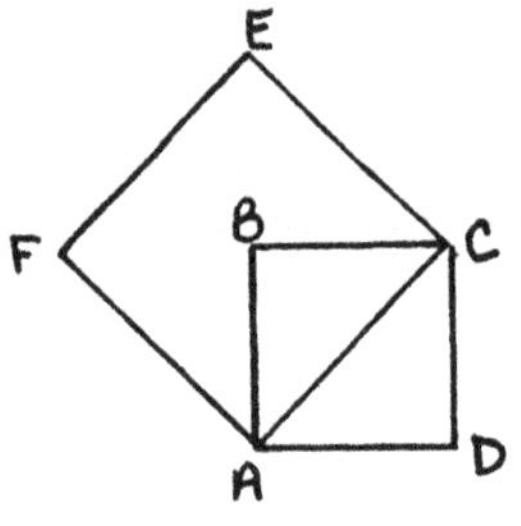

9. Construct a square whose area is equal to the sum of the areas of two given squares. (Hint: Use the Pythagorean theorem and construct the figure.)

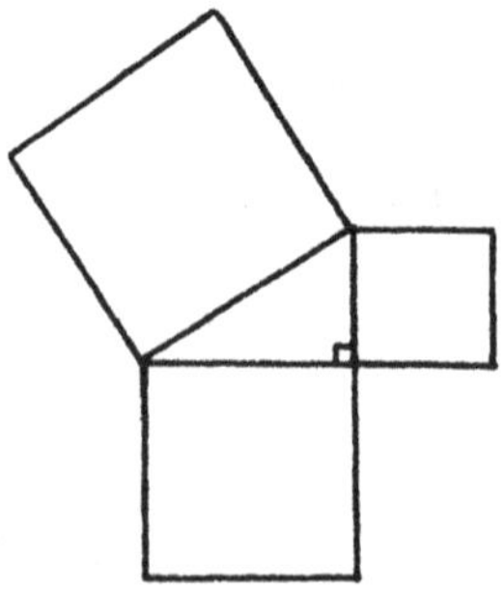

10. Construct a square with area twice that of a given square. (Hint: See problem 9.)

11. Given: Point P anywhere within equilateral $\triangle ABC$. $\overline{PL}$, $\overline{PM}$, and $\overline{PN}$ are perpendicular to their sides.

Prove: $m(\overline{PL}) + m(\overline{PM}) + m(\overline{PN})$ equals the measure of the altitude $\overline{CQ}$, of $\triangle ABC$. (Hint: The total area is the sum of areas of $\triangle APC$, $\triangle APB$, and $\triangle BPC$.)

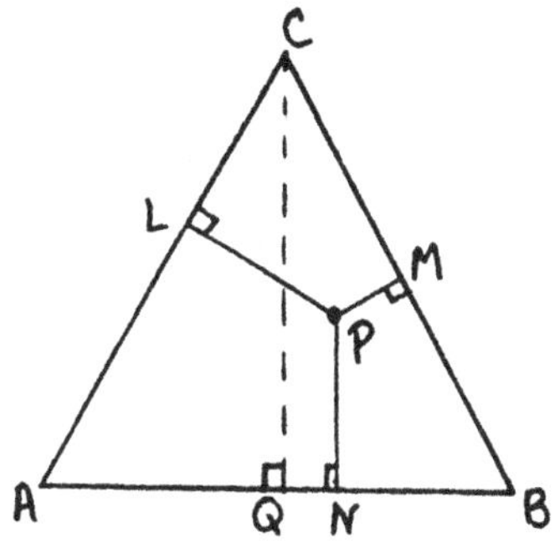

12. Two spheres, one with a radius of 6 cm and one with radius of 3 cm, have the same center. Find the volume of space contained between their surfaces. (252π cubic cm)

13. Find the volume of space that is contained in the cylinder but not the cone. (108π cu in.)

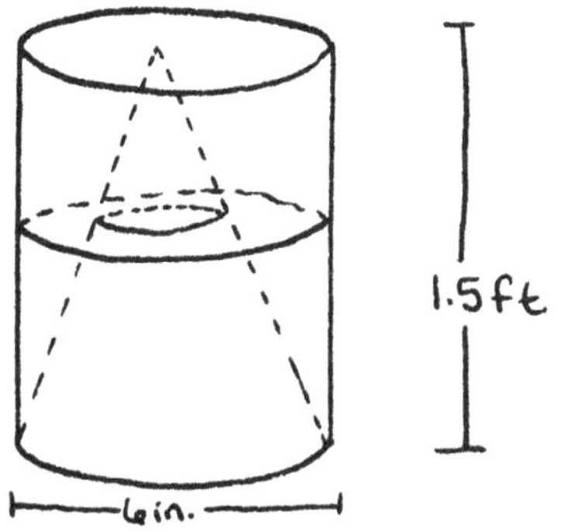

LOGO EXERCISES

In exercises 1 – 4 write a Logo program to draw the following figures with the given angles. Remember to use the properties of the lengths of side of the 30° – 60° and the 45° – 45° right triangles to find decimal approximations for the distances.

1.

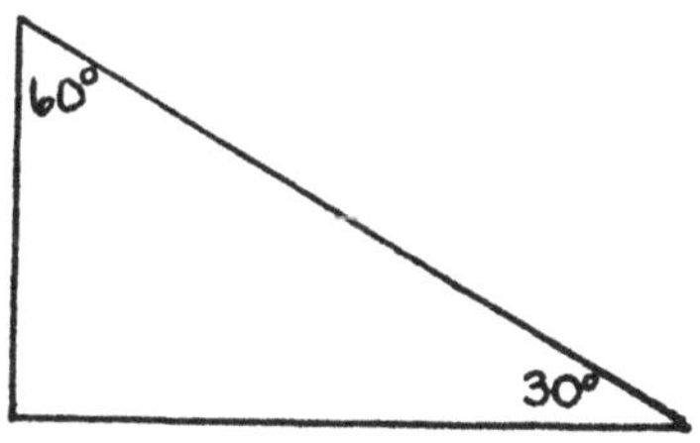

2.

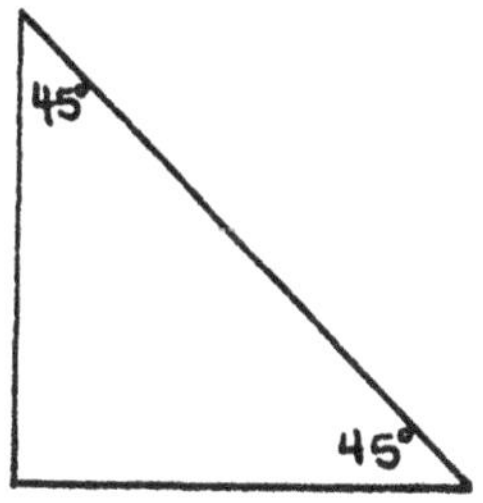

3.

4.

Chapter 5 Test

Form 1

Decide if each of the following statements is true or is false. Then mark T or F in the space provided.

_________ 1. The area of a parallelogram with a base measuring 6 in. and a height of 8 in. is 48 sq in.

_________ 2. The area of $\triangle ABC$ is 24 sq cm.

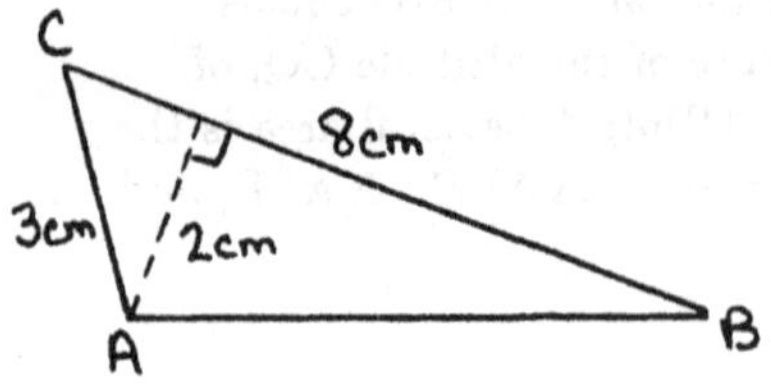

_________ 3. The area of a trapezoid with bases 10 m and 12 m long and a height of 6 m is 132 sq m.

_________ 4. The area of a square with a diagonal $\sqrt{2}$ in. long is 4 sq in.

_________ 5. A rhombus has diagonals 4 cm and 8 cm long. The area of the rhombus is 16 sq cm.

_________ 6. The area of a circle with a diameter measuring 14 m is 49π sq m.

_________ 7. A triangle with sides measuring 6 ft, 8 ft and 10 ft is a right triangle.

_________ 8. In a $30^\circ - 60^\circ$ right triangle, the side opposite the 60° angle is twice as long as the side opposite the 30° angle.

_________ 9. The area of an equilateral triangle with an altitude $6\sqrt{3}$ cm long is 18 sq cm.

_________10. The volume of a rectangular solid measuring 4 ft by 12 ft by 6 ft is 48 cubic ft.

Solve each problem and place the answer in the space provided.

_________11. Find m($\overline{AC}$).

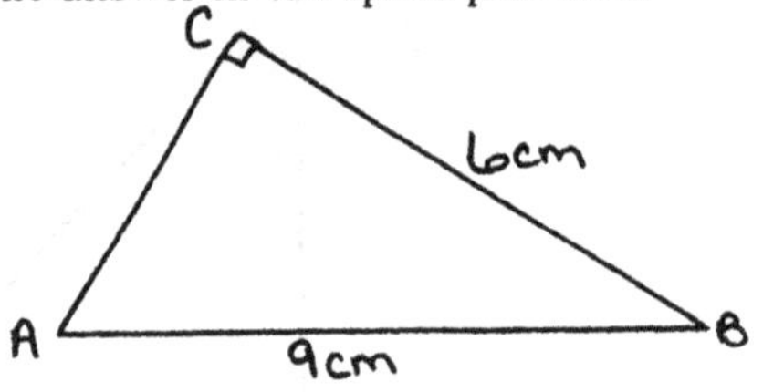

_________12. Find the height of a trapezoid if its bases are 8 in. and 14 in. long and its area is 66 sq in.

_________13. A triangle has a base 9 ft long and a height of 8 ft. Find the side of a square that has an area equal to that of the triangle.

_________14. Find the area of the figure if $m(\overline{HG}) = m(\overline{DC})$, and $m(\overline{HA}) = m(\overline{CB})$.

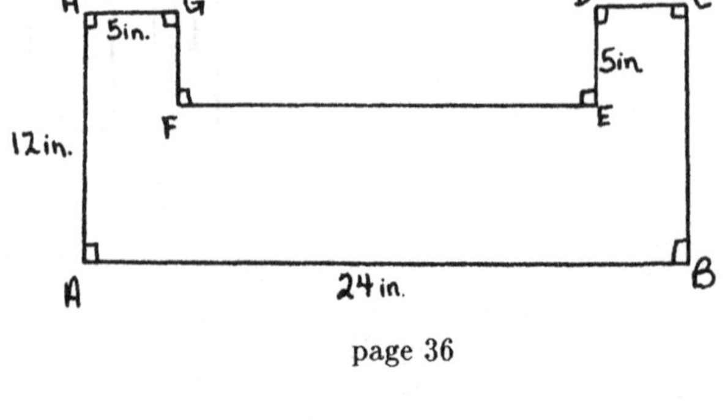

________15. A square has a diagonal that is 5 yd long. What is its area?

________16. A square has sides that are 8 cm long. Find the radius of a circle that has an area equal to that of the square.

________17. Find the area of an equilateral triangle with an altitude that is $10\sqrt{3}$ in. long.

________18. In the figure, $\overline{BC} \cong \overline{AC}$ and $m(\overline{AB}) = 10$ft. Find the area of $\triangle ABC$.

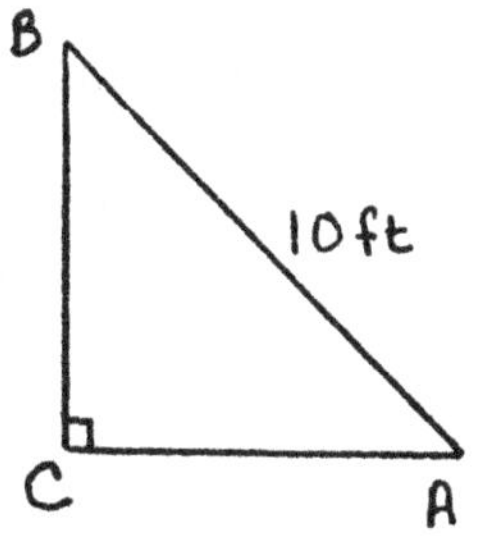

________19. If two legs of a right triangle are 10 m and 24 m long, determine the length of the hypotenuse.

________20. A rectangle has sides that are 32 cm and 2 cm long. Find the side of a square that has an area equal to that of the rectangle.

________21. Find the surface area, excluding the bases, of a right-circular cylinder whose base has a diameter measuring 14 ft and whose height is 15 ft.

________22. Find the volume of a right-circular cone with a base of radius 5 cm and a height of 12 cm.

23. Given: Two concentric circles with common center O with radii R and r.

Prove: The area between the circles equals $\pi(R^2 - r^2)$.

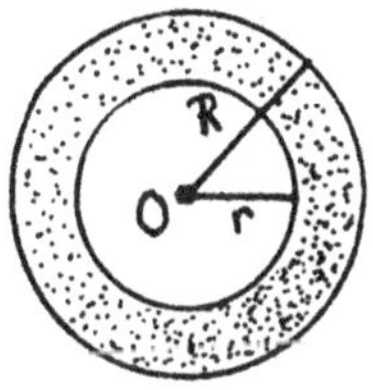

STATEMENTS	REASONS

Chapter 5 Test

Form 2

Decide if each of the following statements is true or is false. Then mark T or F in the space provided.

_________ 1. The area of a rectangle with sides measuring 7 in. and 8 in. is 28 sq in.

_________ 2. If two figures have the same area, then they are congruent.

_________ 3. The area of $\triangle ABC$ is 26 sq cm.

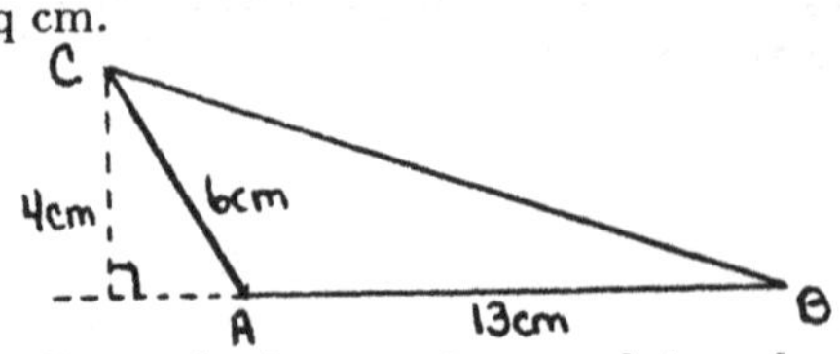

_________ 4. The area of a rhombus with diagonals that are 6 cm and 4 cm long is 24 sq cm.

_________ 5. The area of a square with a diagonal that is $\sqrt{6}$ in long is 3 sq in.

_________ 6. The area of a circle with a diameter that is 10 cm long is 25π sq cm.

_________ 7. A triangle with sides measuring 10 cm, 24 cm and 26 cm is a right triangle.

_________ 8. In the $30^\circ - 60^\circ$ right triangle below, the side opposite the 30° angle is $3\sqrt{3}$ cm long.

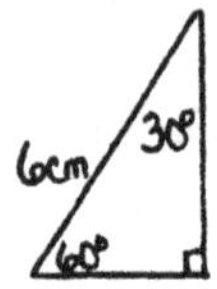

_________ 9. If the legs of an isosceles right triangle have measure of $6\sqrt{2}$ cm, then the hypotenuse has length of 12 cm.

_________ 10. The circumference of a circle with radius that is 3 cm long is 6π cm.

Solve each problem and place the answer in the space provided.

_________ 11. If the altitude of a parallelogram has length of 8 cm and the base has length of 12 cm, find the area.

_________ 12. Find m($\overline{AC}$).

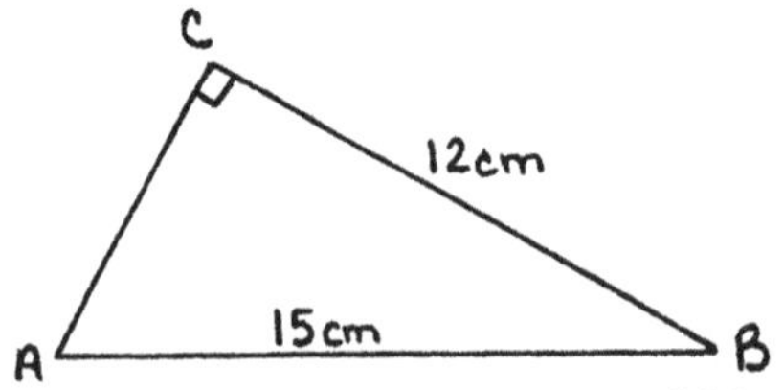

_________ 13. Find the height of the trapezoid with an area of 36 sq cm and bases with lengths of 16 cm and 20 cm.

_________ 14. Find the area of the following figure if $\overline{AB} \parallel \overline{EF}$, $\overline{EF} \parallel \overline{DC}$, and $\overline{EF} \cong \overline{DC}$.

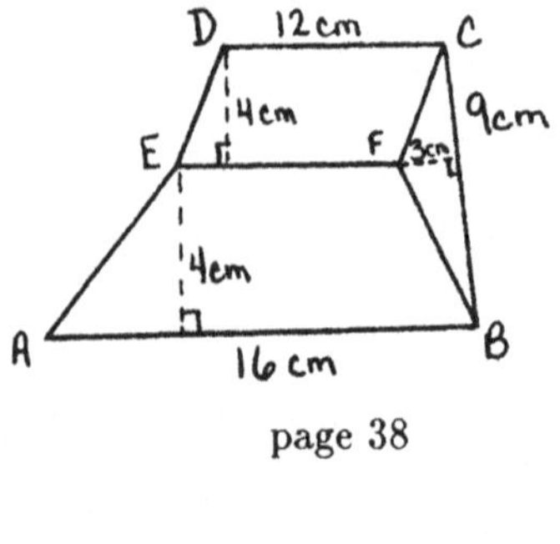

________ 15. Find the area of a square which has diagonals of length 8 yds.

________ 16. A square has a side measuring 10 cm. Find the length of the radius of a circle that has an area equal to that of the square.

________ 17. Find the area of an equilateral triangle with an altitude of length $8\sqrt{3}$ in.

________ 18. Find the area of trapezoid ABCD.

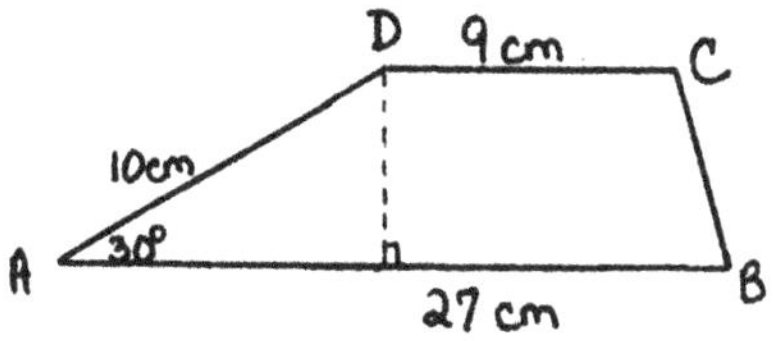

________ 19. If the length of the hypotenuse of a right triangle is 26 cm and one of the legs has length of 10 cm, find the length of the other leg.

________ 20. A triangle has a base of length 8 m and a height of 4 m. Find the length of the side of a square that has an area equal to that of the triangle.

________ 21. Find the surface area, excluding the bases, of a right-circular cylinder whose base has a radius of 4 m and whose height is 9 m.

________ 22. Find the volume of a right prism whose height is 7 cm and whose base is a right triangle with legs that are 5 cm and 6 cm long.

23. Given: Square ABCD inscribed in circle O with common centers at O.

Prove: The area inside the circle and outside the square equals $(\pi - 2)r^2$.

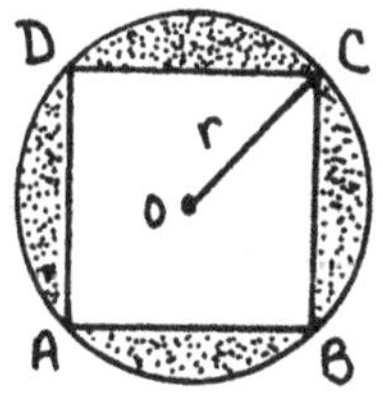

STATEMENTS	REASONS

Chapter 5 Test

Form 3

Decide if each of the following statements is true or is false. Then mark T or F in the space provided.

_________ 1. If three sides of one triangle are congruent to three sides of a second triangle, then the areas are equal.

_________ 2. The area of a rectangle with sides measuring 6 in. and 7 in. is 42 sq in.

_________ 3. The area of $\triangle ABC$ is 28 sq cm.

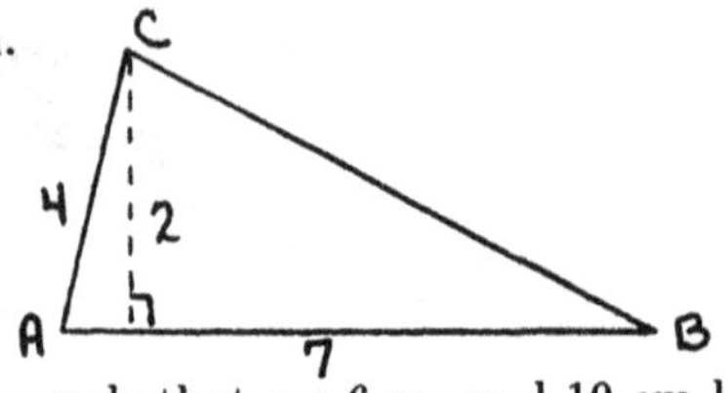

_________ 4. The area of a rhombus with diagonals that are 6 cm and 10 cm long is 30 sq cm.

_________ 5. The area of a square with a diagonal that is $\sqrt{10}$ in long is 5 sq in.

_________ 6. The circumference of a circle with a diameter that is 7 m long is 7π m.

_________ 7. A triangle with sides measuring 12 ft, 16 ft and 20 ft is a right triangle.

_________ 8. In the $30° - 60°$ right triangle, the side opposite the $60°$ angle is $\sqrt{3}$ times the hypotenuse.

_________ 9. If the hypotenuse of an isosceles right triangle is $4\sqrt{2}$ cm long, then the legs have length of 4 cm.

_________ 10. The area of a circle with a diameter that is 10 in. long is 25π sq in.

Solve each problem and place the answer in the space provided.

_________ 11. If the altitude of a parallelogram has length of 6 m and the base has length of 11 m, find the area.

_________ 12. Find m($\overline{BC}$).

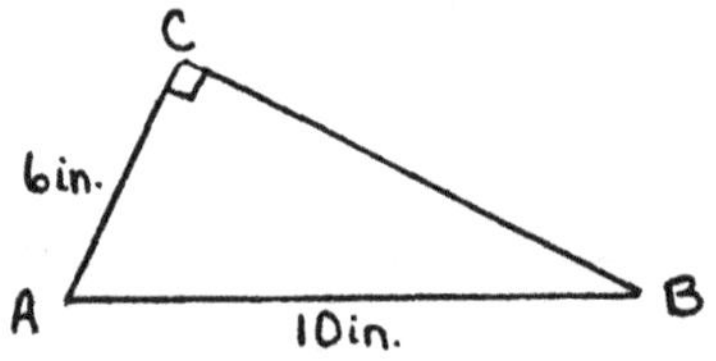

_________ 13. Find the height of the trapezoid with an area of 11 sq cm and bases with lengths of 4 cm and 7 cm.

_________ 14. Find the area of the following figure.

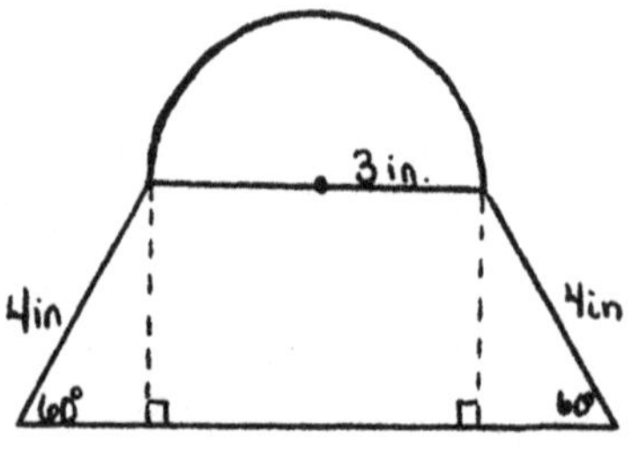

________15. Find the area of a rhombus which has diagonals of length 6 cm and 9 cm.

________16. A circle has an area of 49π sq in. Find the length of the radius of the circle.

________17. Find the area of an equilateral triangle with an altitude of length $6\sqrt{3}$ ft.

________18. In the figure, $\overline{BC} \cong \overline{AC}$ and $m(\overline{AB}) = 12\sqrt{2}$ yd. Find the area of $\triangle ABC$.

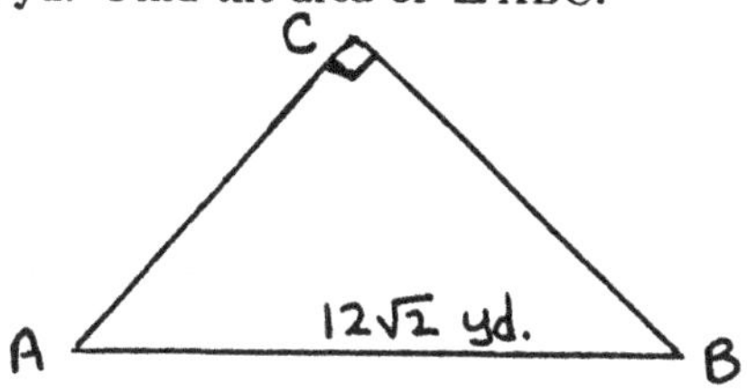

________19. If the length of the hypotenuse of a right triangle is 20 in. and one of the legs has length of 12 in., find the length of the other leg.

________20. A rhombus has diagonals of lengths of 1 m and 8 m. Find the length of the side of the square that has an area equal to that of the rhombus.

________21. Find the surface area, excluding the bases, of a right-circular cylinder whose base has a radius of 3 cm and whose height is 7 cm.

________22. Find the volume of a cone with a diameter that is 8 cm long and a height of 10 cm.

23. Given: Square ABCD with sides of length s, and circle O inscribed in ABCD with common centers at O.

Prove: The area inside the square and outside the circle equals $\left(1 - \frac{\pi}{4}\right)s^2$.

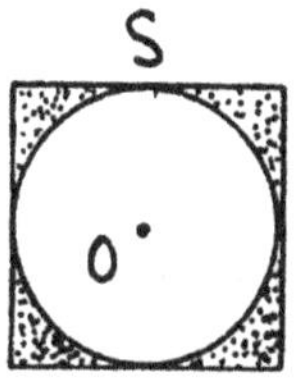

STATEMENTS	REASONS

CHAPTER 6

RATIO, PROPORTION, AND SIMILARITY

SUPPLEMENTARY EXERCISES

1. Prove: The ratio of the areas of two similar triangles is the square of the ratio of a pair of corresponding sides.

2. Given: $\triangle ABC$ with altitudes $\overline{AE}$ and $\overline{BD}$ intersecting at F.

Prove: $\triangle ADF \sim \triangle BEF$.

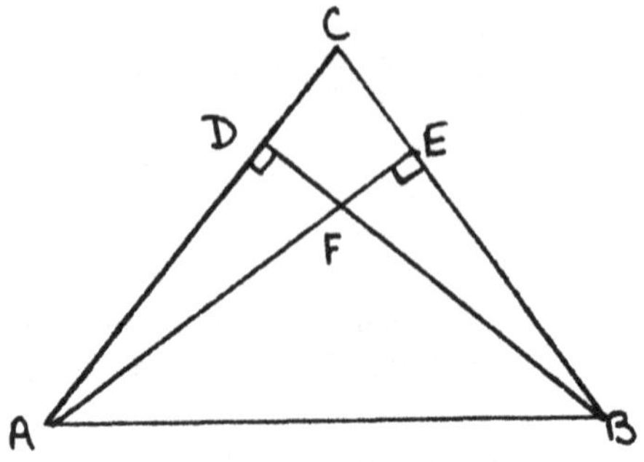

3. Given: Trapezoid ABCD with diagonals $\overline{AC}$ and $\overline{BD}$ intersecting at E.

Prove: $\dfrac{m(\overline{AE})}{m(\overline{EC})} = \dfrac{m(\overline{BE})}{m(\overline{ED})}$.

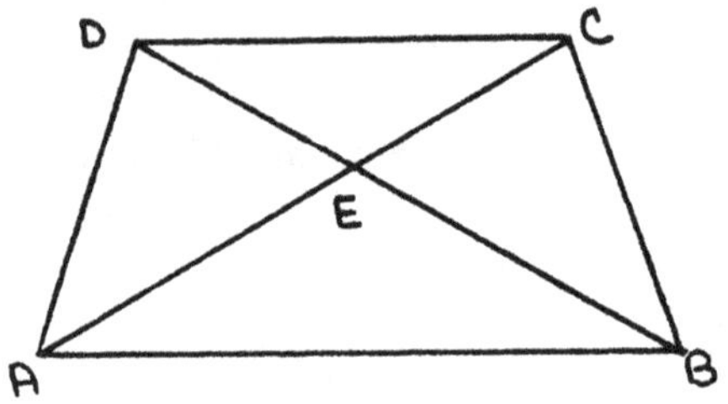

4. Given right $\triangle ABC$ with $\overline{CF} \perp \overline{AB}$. Find the length of segment $\overline{CF}$. (6)

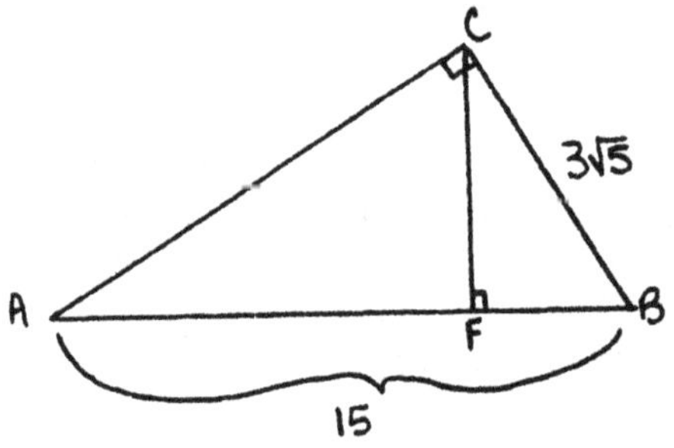

5. Given: Right $\triangle ABC$ with right $\angle C$. $\overline{DG} \perp \overline{AB}$ and $\overline{EF} \perp \overline{AB}$.

Prove: $\triangle ADG \sim \triangle BEF \sim \triangle ABC$.

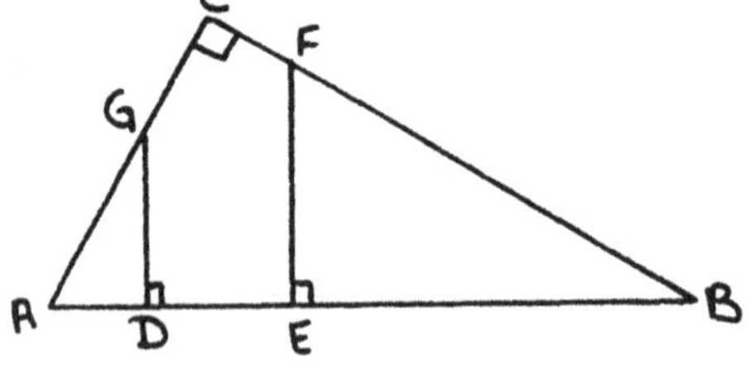

6. Find the value of x in each of the following:

a)

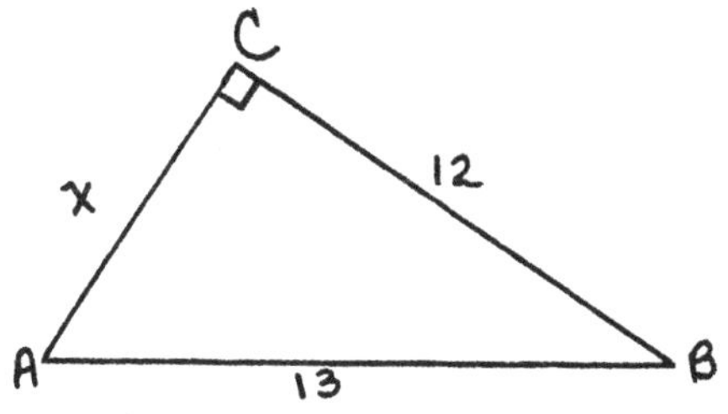

$(x=5)$

b)

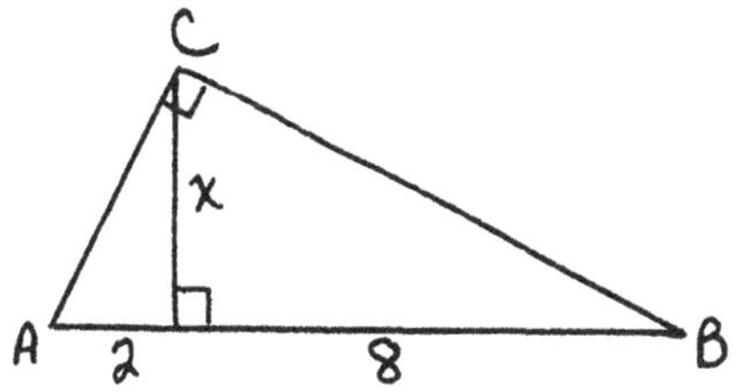

$(x=4)$

c)

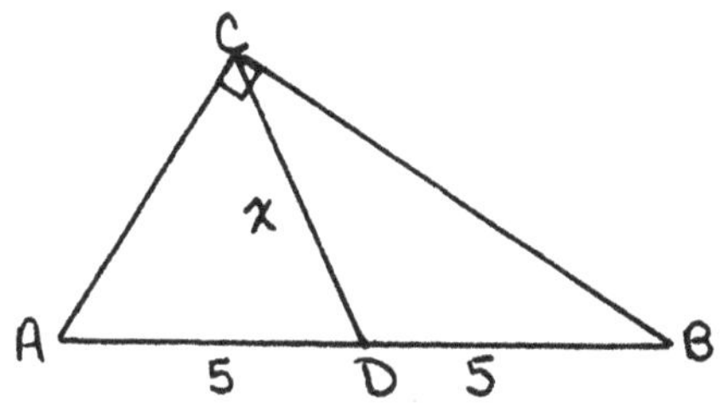

$(x=5)$

d)

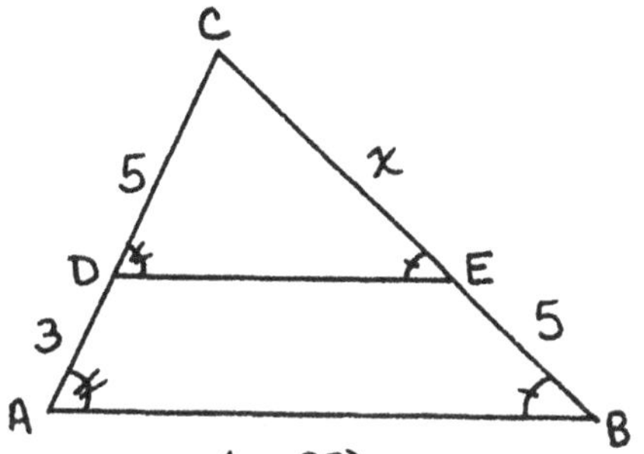

$(x=\frac{25}{3})$

e)

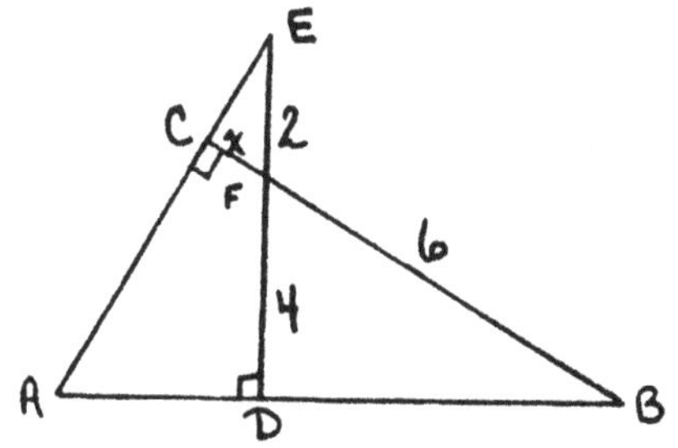

$(x=\frac{4}{3})$

f)

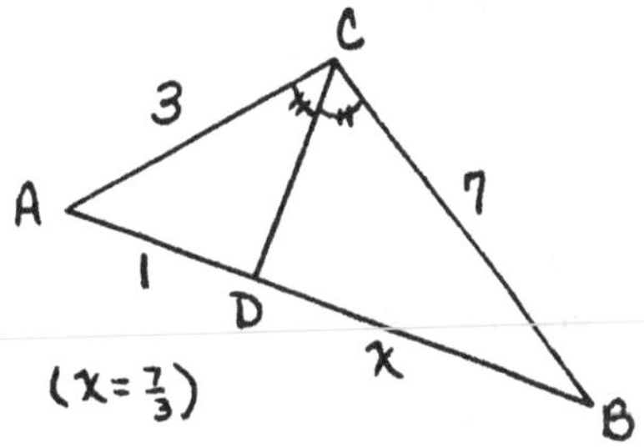

$(x=\frac{7}{3})$

g)

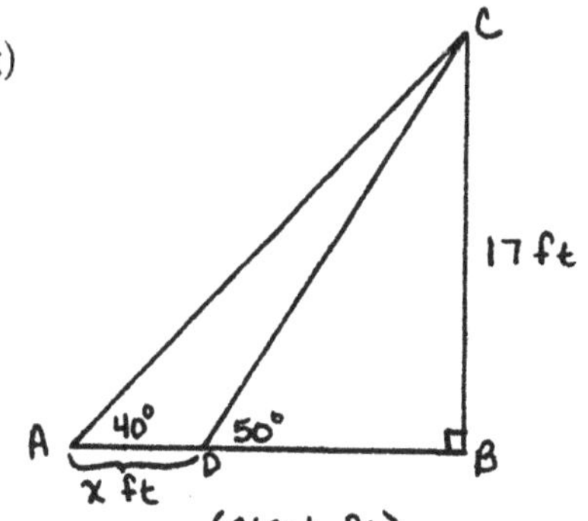

$(x \approx 6\text{ft})$

h)

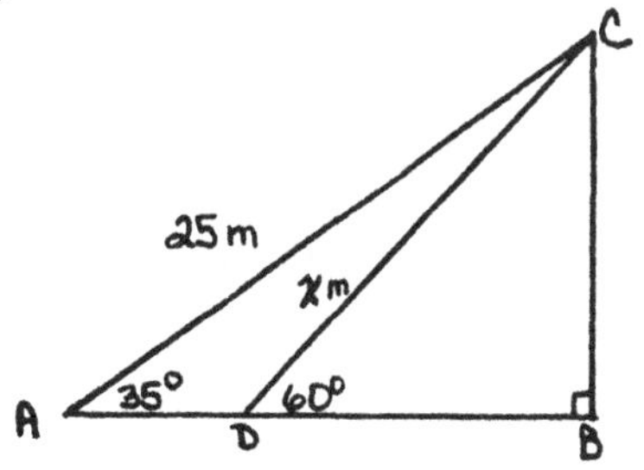

$(x \approx 16.6\text{m})$

i)

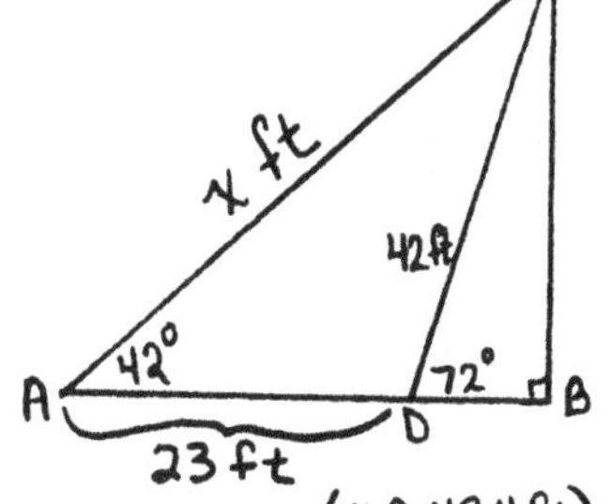

$(x \approx 48.4\text{ft})$

j)

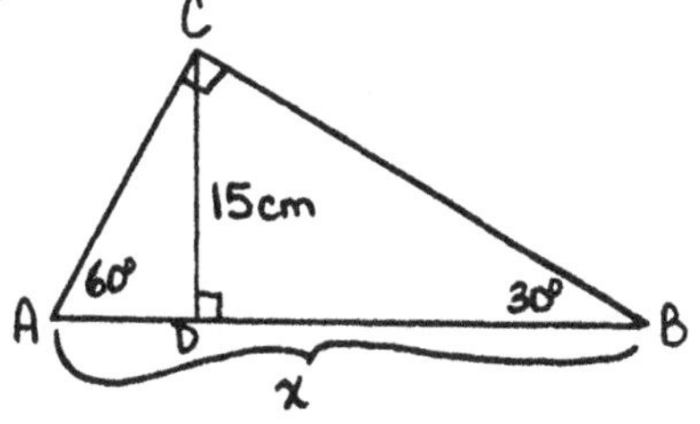

$(x \approx 34.6\text{ cm})$

Chapter 6 Test

Form 1

Decide if each of the following statements is true or is false. Then mark T or F in the space provided.

_________ 1. A ratio is the comparison of two numbers by their indicated quotient.

_________ 2. A proportion is a ratio.

_________ 3. The fourth proportional of 6, 12, and 18 is 24.

_________ 4. In a proportion, the product of the means is equal to the product of the extremes.

_________ 5. A line perpendicular to one side of a triangle divides the other two sides proportionally.

_________ 6. The measure of side $\overline{BC}$ is 18.

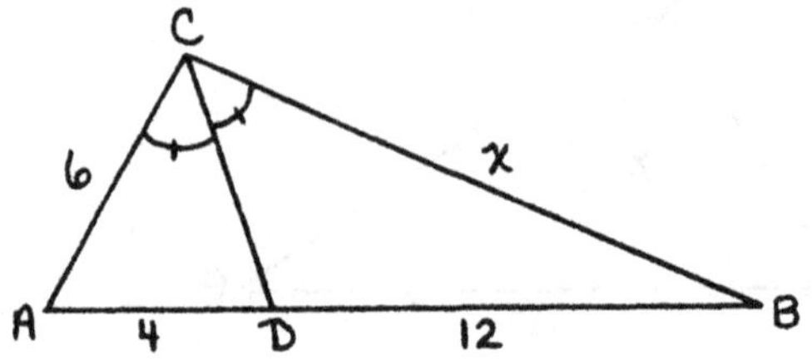

_________ 7. All similar triangles are congruent.

_________ 8. If two right triangles have a respective pair of acute angles congruent, the triangles are similar.

_________ 9. The altitude upon the base of a triangle is the mean proportional between the segments of the base of the triangle.

_________ 10. All squares are similar.

_________ 11. The sine ratio of an acute angle of a right triangle is the ratio of the length of the side opposite the angle to the length of the hypotenuse.

_________ 12. If $\cos A = \frac{3}{5}$, then $\sin A = \frac{4}{5}$.

Solve each problem and place the answer in the space provided.

_________ 13. Find the positive mean proportional between 2 and 32.

_________ 14. Solve the proportion $\frac{x+6}{2} = \frac{4}{1}$ for x.

_________ 15. Find x if $m(\overline{AC}) = 6$.

_________ 16. If $x = 16$, find $m(\overline{CD})$.

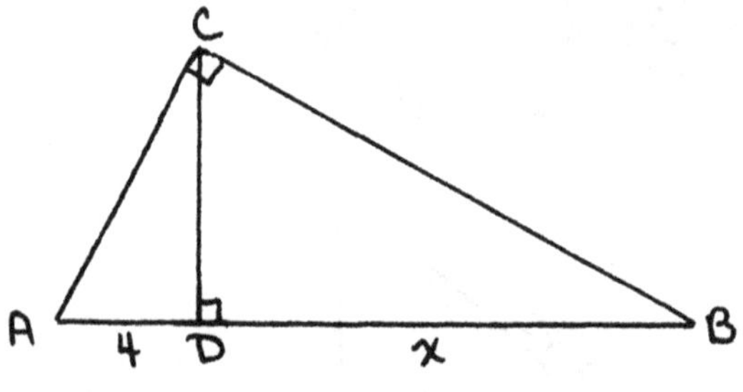

(Questions 15-16)

________17. $\overline{DE} \parallel \overline{AB}$ and $m(\overline{AC}) = 16$.

Find $m(\overline{DC})$ if $m(\overline{CE}) = 2$ and $m(\overline{BE}) = 8$.

________18. $\overline{DE} \parallel \overline{AB}$, $m(\overline{AC}) = 16$ and $m(\overline{BC}) = 24$.

Find $m(\overline{AD})$ if $m(\overline{CE}) = 6$.

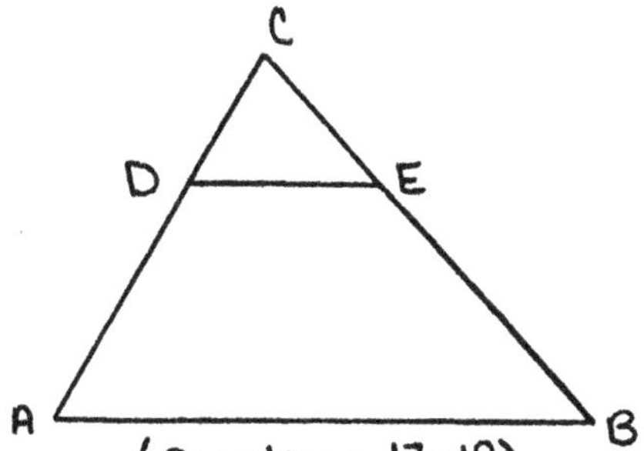

(Questions 17-18)

________19. In the figure, $\triangle ABC \sim \triangle DEF$.

Find $m(\overline{DE})$.

________20. In the figure, $\triangle ABC \sim \triangle DEF$.

Find $m(\overline{EF})$.

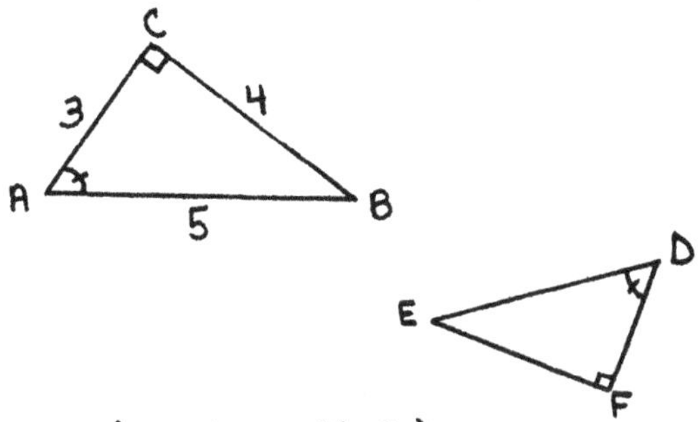

(Questions 19-20)

________21. One triangle has sides measuring 3, 4, and 6. Another triangle has sides measuring 15, 20, and x. Find x so that the triangles will be similar.

________22. A woman standing on level ground views the top of a building at an angle of 73°. How tall is the building if she is 15 ft from its base?

________23. If $\sin A = \frac{12}{13}$, find $\cos A$.

24. Given: $\overline{DC}$ and $\overline{AB}$ intersect at E.

$\overline{AD} \perp \overline{AB}$ and $\overline{BC} \perp \overline{AB}$.

Prove: $\dfrac{m(\overline{AD})}{m(\overline{BC})} = \dfrac{m(\overline{AE})}{m(\overline{BE})}$.

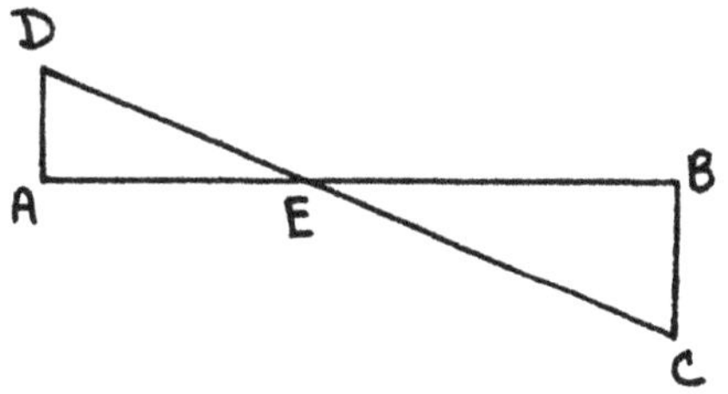

STATEMENTS	REASONS

Chapter 6 Test

Form 2

Decide if each of the following statements is true or is false. Then mark T or F in the space provided.

_________ 1. If $\frac{a}{b} = \frac{c}{d}$, then $\frac{a-b}{b} = \frac{c-d}{d}$.

_________ 2. In a proportion, the denominators of the two ratios are equal.

_________ 3. The numbers 2, 5, 8, and 20 form a proportion.

_________ 4. All isosceles right triangles are similar.

_________ 5. The measures of all corresponding sides of similar triangles are in proportion.

_________ 6. The cosine of an acute angle of a right triangle is the ratio of the length of the side opposite the angle to the length of the hypotenuse.

_________ 7. The length of the altitude on the hypotenuse of a right triangle is the mean proportional between the lengths of the segments of the hypotenuse.

_________ 8. All regular pentagons are similar.

_________ 9. A line parallel to one side of a triangle divides the other two sides into congruent segments.

_________ 10. All parallelograms are similar.

_________ 11. If an acute angle of one right triangle is complementary to an acute angle of another right triangle, then the triangles are similar.

_________ 12. It is possible for cos A to equal sin A.

Solve each problem and place the answer in the space provided.

_________ 13. Find the positive mean proportional between 2 and 32.

_________ 14. Solve the proportion $\frac{x-2}{2} = \frac{x}{4}$ for x.

_________ 15. Find x if $m(\overline{AC}) = 4$.

_________ 16. If $x = 18$, find $m(\overline{CD})$.

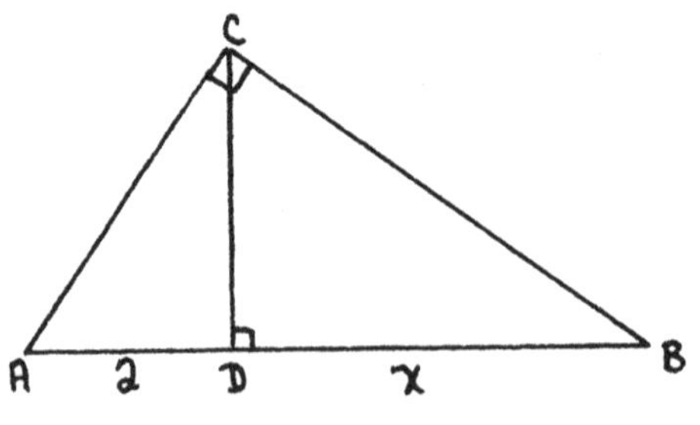

(Questions 15-16)

________17. $\overline{DE} \parallel \overline{AB}$ and $m(\overline{AC}) = 9$.

Find $m(\overline{DC})$ if $m(\overline{CE}) = 14$ and $m(\overline{BE}) = 4$.

________18. $\overline{DE} \parallel \overline{AB}$, $m(\overline{AC}) = 11$ and $m(\overline{BC}) = 33$.

Find $m(\overline{AD})$ if $m(\overline{CE}) = 18$.

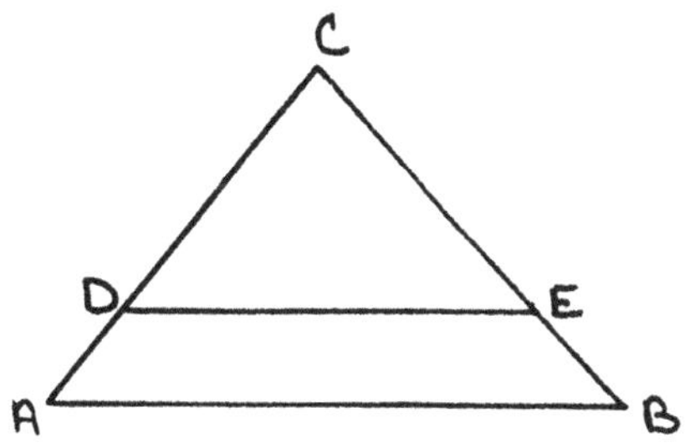

(Questions 17-18)

________19. In the figure, $\triangle ABC \sim \triangle DEF$.

Find $m(\overline{EF})$.

________20. In the figure, $\triangle ABC \sim \triangle DEF$.

Find $m(\overline{DF})$.

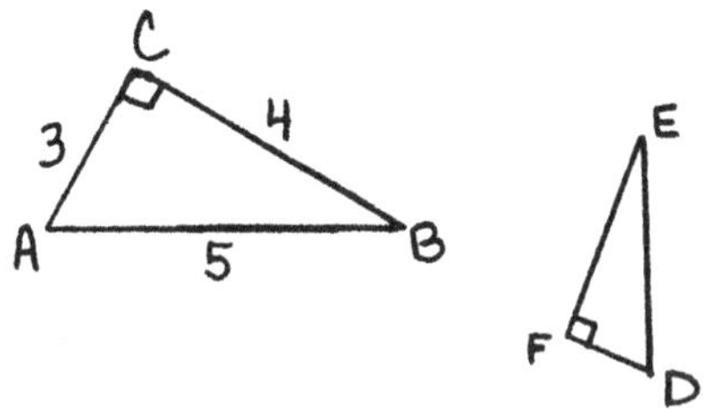

(Questions 19-20)

________21. One triangle has sides measuring 2, 6, and 7. Another triangle has sides measuring 14, 42, and x. Find x so that the triangles will be similar.

________22. A man standing on level ground views the top of a flagpole at an angle of 37.5°. How tall is the flagpole if he is 23.0 m from its base?

________23. If $\sin A = \frac{12}{13}$, find $\tan A$.

24. Given: $\overline{AC} \cong \overline{AE}$.

$\angle CBD \cong \angle EFD$.

Prove: $\triangle BCD \sim \triangle FED$

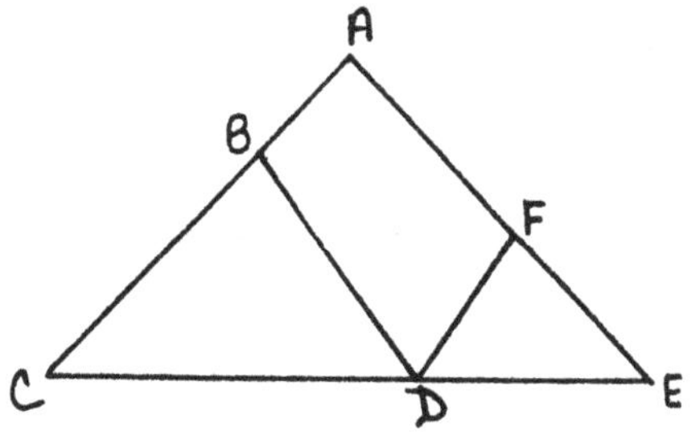

STATEMENTS	REASONS

Chapter 6 Test

Form 3

Decide if each of the following statements is true or is false. Then mark T or F in the space provided.

________ 1. If $\frac{a}{b} = \frac{c}{d}$, then $\frac{c}{a} = \frac{d}{b}$.

________ 2. The fourth proportional of 5, 11, and 13 is 18.

________ 3. If $\frac{a}{b} = \frac{c}{d}$, then $\frac{a}{c} = \frac{d}{b}$.

________ 4. A line parallel to one side of a triangle forms a second trinagle similar to the first.

________ 5. The measure of side $\overline{DB}$ is 5.

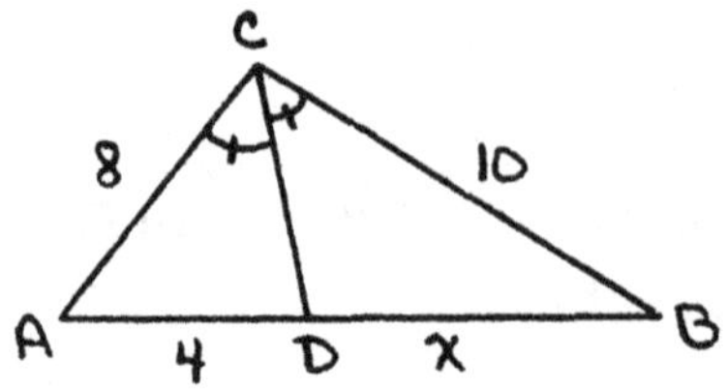

________ 6. All pentagons are similar.

________ 7. If two sides and an included angle of one triangle are congruent to two sides and an included angle of another triangle, the triangles are similar.

________ 8. Sides of similar triangles have equal measure.

________ 9. All equiangular triangles are similar.

________10. In $\triangle ABC$, $\cos C = \frac{5}{13}$.

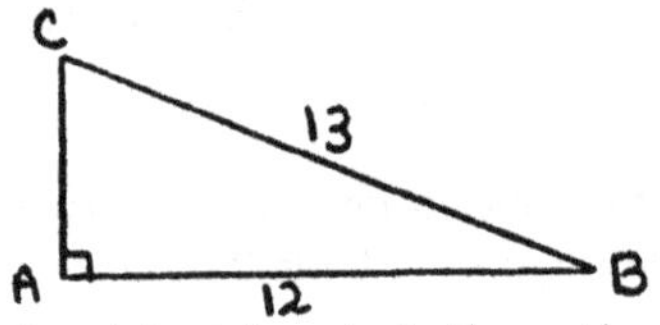

________11. The tangent ratio of an acute angle of a right triangle is the ratio of the length of the opposite side of the angle to the length of the side adjacent to the angle.

________12. If $\cos A = \frac{\sqrt{2}}{2}$ then $\tan A = 1$.

Solve each problem and place the answer in the space provided.

________13. Find the positive mean proportional between 4 and 25.

________14. Solve the proportion $\frac{x+5}{6} = \frac{5}{1}$ for x.

________15. Find x if $m(\overline{AD}) = 5$ and $m(\overline{CD}) = 10$.

________16. If $m(\overline{AD}) = 3$ and $m(\overline{AB}) = 15$, find $m(\overline{CD})$.

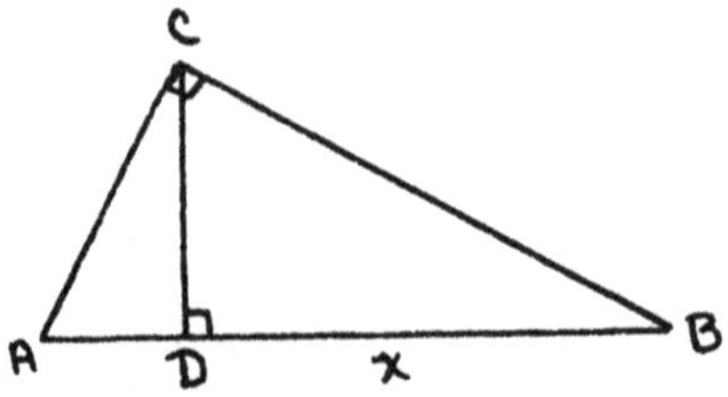

(Questions 15-16)

________17. $\overline{DE} \parallel \overline{AB}$ and $m(\overline{CD}) = 6$.

Find $m(\overline{AB})$ if $m(\overline{DA}) = 2$ and $m(\overline{DE}) = 9$.

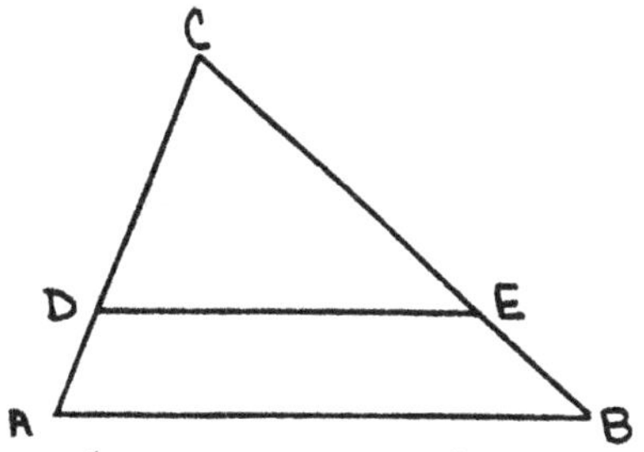

________18. $\overline{DE} \parallel \overline{AB}$, $m(\overline{DC}) = 4$ and $m(\overline{AC}) = 6$.

Find $m(\overline{EB})$ if $m(\overline{CE}) = 7$.

(Questions 17-18)

________19. In the figure, $\overline{DE} \parallel \overline{AB}$.

Find the perimeter of $\triangle ABC$.

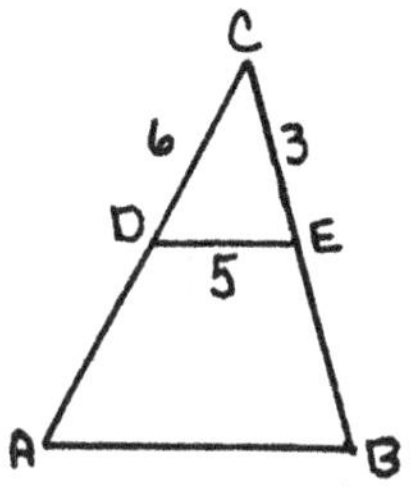

(Question 19)

________20. If $\cos A = \frac{12}{13}$, find tan A.

________21. The distance from a point on the ground to the top of a tower is 21 ft. How tall is the tower if the angle formed by looking at its top is 59°?

________22. Find sin 30°.

________23. If $\cos A = \frac{12}{13}$, find sin A.

24. Given: $\angle B \cong \angle ADE$.

Prove: $\frac{m(\overline{EA})}{m(\overline{AC})} = \frac{m(\overline{DE})}{m(\overline{BC})}$.

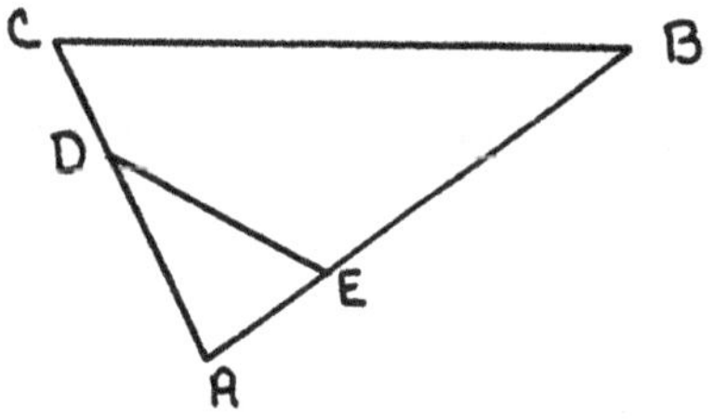

STATEMENTS	REASONS

CHAPTER 7

CIRCLES AND MORE ON SIMILARITY

SUPPLEMENTARY EXERCISES

1. Prove that a trapezoid inscribed in a circle is isosceles.

2. Given: Line $\overline{PR}$ tangent to circle O at point Q.
   $\overline{AB}$ is a diameter of circle O.
   $\overline{AP} \perp \overline{PR}$ and $\overline{BR} \perp \overline{PR}$.

   Prove: $m(\overline{AP}) + m(\overline{BR}) = m(\overline{AB})$.
   (Hint: Draw $\overline{OQ}$.)

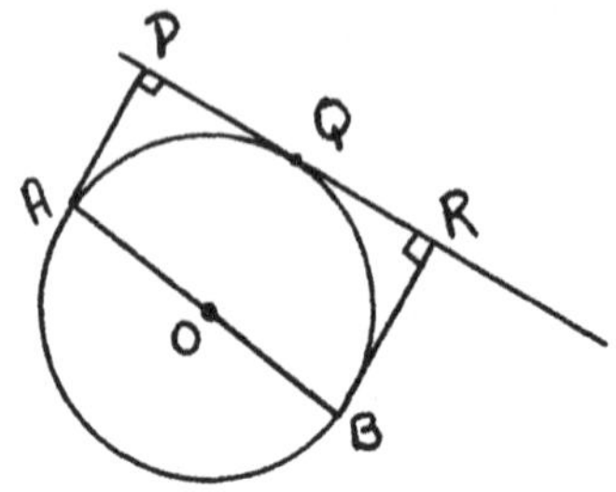

3. Prove: If two circles are tangent externally, the common external tangent is bisected by their common internal tangent.

4. Given: Secant $\overline{ABCD}$, and diameter $\overline{LM}$ of circle O.
   $\overline{LA} \perp \overline{AD}$ and $\overline{MD} \perp \overline{AD}$.

   Prove: $\overline{AB} \cong \overline{CD}$.
   (Hint: Construct $\overline{OQ} \perp \overline{AD}$.)

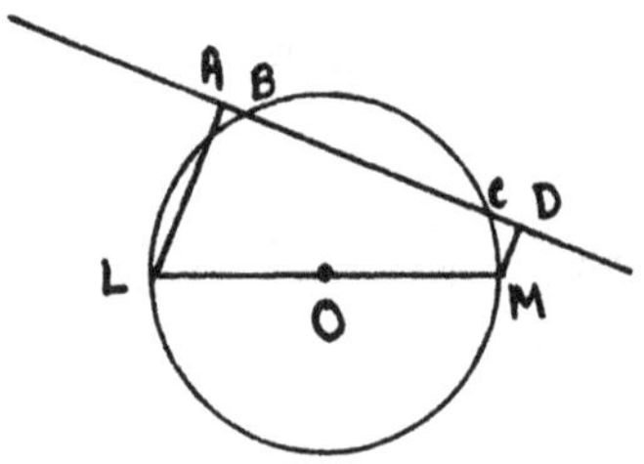

5. Given: Circle O with tangents $\overline{AB}$, $\overline{CD}$, and $\overline{BD}$.
   $\overline{AB} \parallel \overline{CD}$.

   Prove: $\angle$BOD is a right angle.

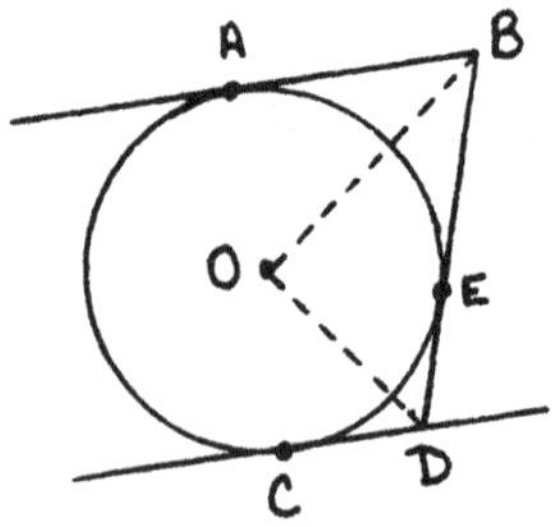

6. Given a rectangle, construct a square with the same area. (Hint: Given a and b, find x such that $ab = x^2$, that is, find the mean proportional between a and b.)

7. Given line l and $\overline{AB}$ intersecting l at C. Construct the circle that passes through A and B and is tangent to line l. (Hint: Theorem 7.19 gives a hint as to the location of point T.)

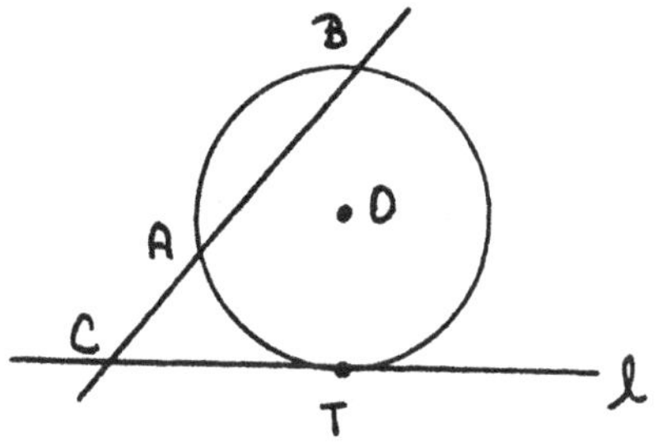

8. Given a rectangle, construct another rectangle with a given width that has an area equal to that of the original rectangle. (Hint: Use Theorem 7.18)

9. Construct a square that has the same area as a given rhombus. (Hint: If a is the length of one diagonal of the rhombus and b is half the length of the other diagonal of the rhombus, then the length of the side, x, of the required square is the mean proportional between a and b.)

10. $\overline{OAB}$ is a diameter. Find:

$m(\angle a) =$ ____________ (ans: 50°)

$m(\angle b) =$ ____________ (ans: 100°)

$m(\widehat{DC}) =$ ____________ (ans: 60°)

$m(\widehat{DB}) =$ ____________ (ans: 100°)

$m(\widehat{DA}) =$ ____________ (ans: 80°)

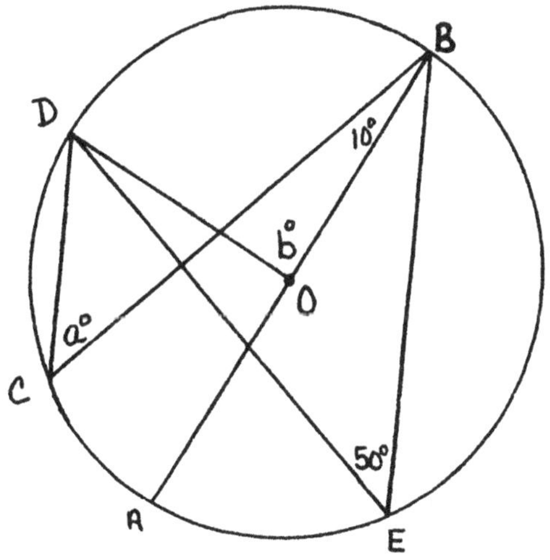

Chapter 7 Test

Form 1

Decide if each of the following statements is true or is false. Then mark T or F in the space provided.

_________ 1. The center of a circle is a point of the circle.

_________ 2. The greatest chord of a circle is the diameter of the circle.

_________ 3. A central angle of a circle has half as many degrees as its intercepted arc.

_________ 4. A line that intersects a circle in exactly one point is called a tangent to the circle.

_________ 5. If an inscribed angle and a central angle of a circle intercept the same arc of the circle, the central angle is twice as large as the inscribed angle.

_________ 6. A bisector of a chord of a circle always passes through the center of the circle.

_________ 7. The opposite angles of an inscribed quadrilateral are supplementary.

_________ 8. If a parallelogram is inscribed within a circle, it must be a rectangle.

_________ 9. An inscribed angle can never be a right angle.

_________10. Two secants can never intersect in a point on a circle.

Solve each problem and place the answer in the space provided.

_________11. Find m($\angle$A).

_________12. Find m($\angle$1).

_________13. Find m($\angle$C).

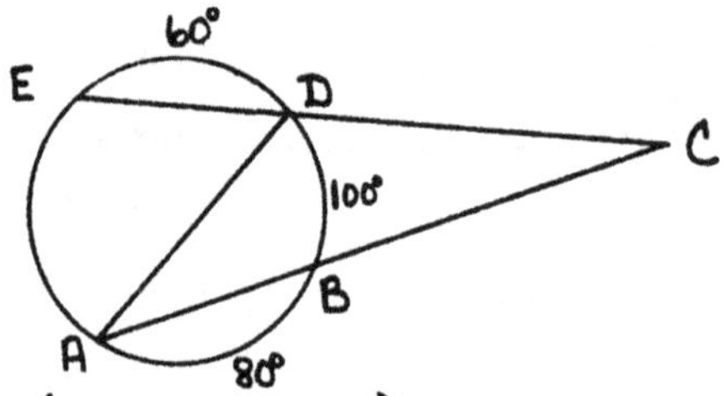

(Questions 11-13)

_________14. Find m($\angle$1).

_________15. Find m($\overline{EB}$) if m($\overline{AE}$) = 8, m($\overline{EC}$) = 4, and m($\overline{DE}$) = 16.

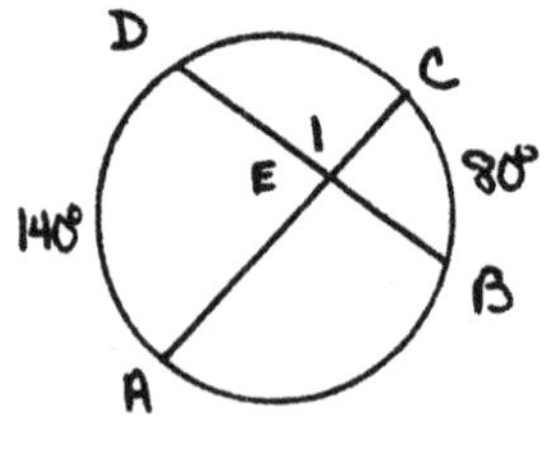

(Questions 14-15)

________16. Find $m(\widehat{AD})$.

________17. Find $m(\angle EDC)$.

________18. Find $m(\angle DCB)$.

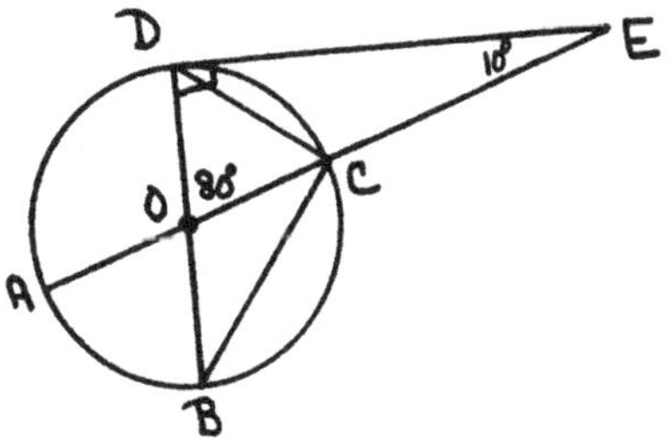

(Questions 16–18)

________19. $\overline{AB} \parallel \overline{DC}$. Find $m(\widehat{DC})$.

________20. $\overline{AB} \parallel \overline{DC}$. Find $m(\widehat{BC})$.

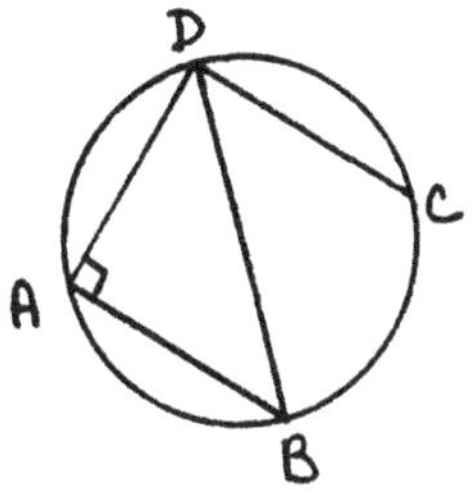

(Questions 19–20)

21. Given: ABCD is an inscribed quadrilateral.
$m(\angle A) = 90°$. $\widehat{AD} \cong \widehat{BC}$.

Prove: ABCD is a parallelogram.

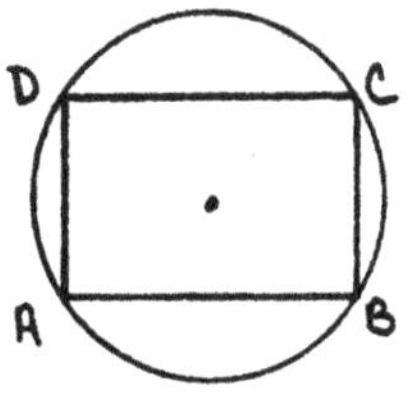

STATEMENTS	REASONS

Chapter 7 Test

Form 2

Decide if each of the following statements is true or is false. Then mark T or F in the space provided.

_________ 1. A radius is a chord of a circle.

_________ 2. A minor arc is an arc that is less than 180°.

_________ 3. If a line is perpendicular to a radius at the point where the radius intersects a circle, then the line is a tangent to the circle.

_________ 4. The measure of an inscribed angle is equal to the measure of its intercepted arc.

_________ 5. The bisector of a chord passes through the center of the circle.

_________ 6. An inscribed angle can never be an obtuse angle.

_________ 7. The triangle formed by an inscribed angle which intersects the endpoints of a diameter is a right triangle.

_________ 8. In a circle, perpendicular lines intersect congruent arcs.

_________ 9. If a quadrilateral is inscribed in a circle, then it must be a rectangle.

_________10. Two tangents can never intersect at a point of a circle.

Solve each problem and place the answer in the space provided.

_________11. Find m(∠CAB).

_________12. Find m($\widehat{AC}$).

_________13. Find m(∠B).

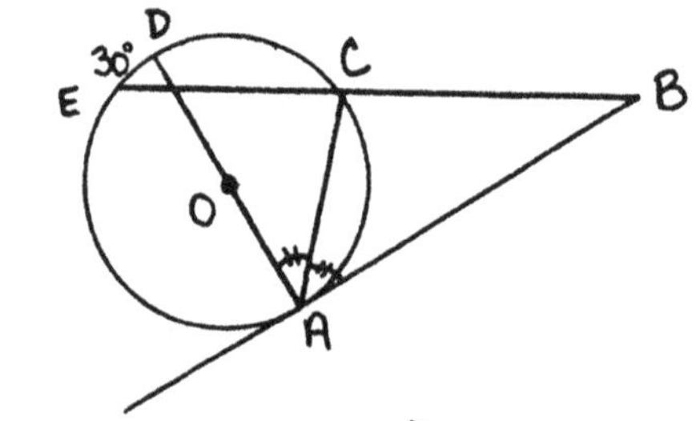

(Questions 11-13)

_________14. Find m(∠1).

_________15. Find m($\overline{EB}$) if m($\overline{AE}$) = 12, m($\overline{AC}$) = 15, and m($\overline{DE}$) = 18.

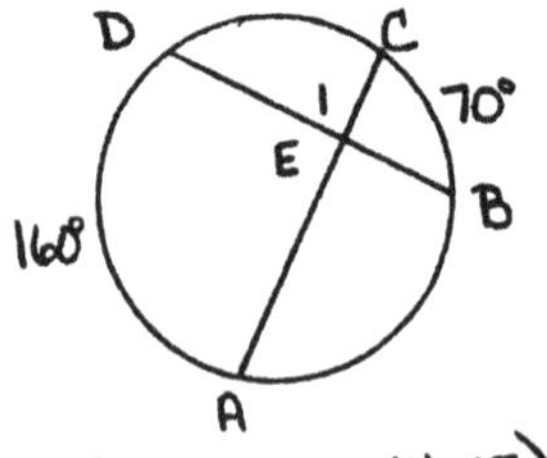

(Questions 14-15)

________16. Find m($\widehat{AB}$).

________17. Find m($\widehat{BC}$).

________18. Find m(∠C).

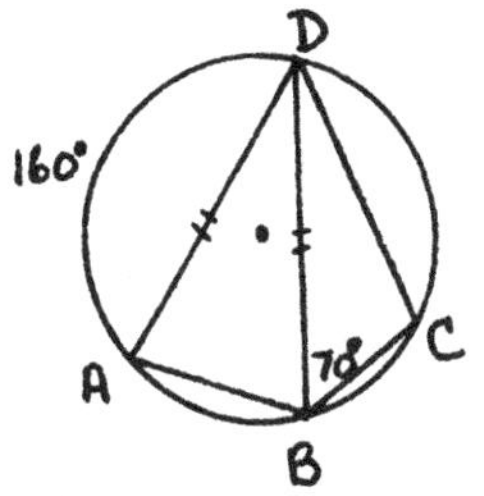

(Questions 16-18)

________19. $\overline{AB} \parallel \overline{DC}$. Find m($\widehat{BC}$).

________20. $\overline{AB} \parallel \overline{DC}$. Find m($\widehat{DC}$).

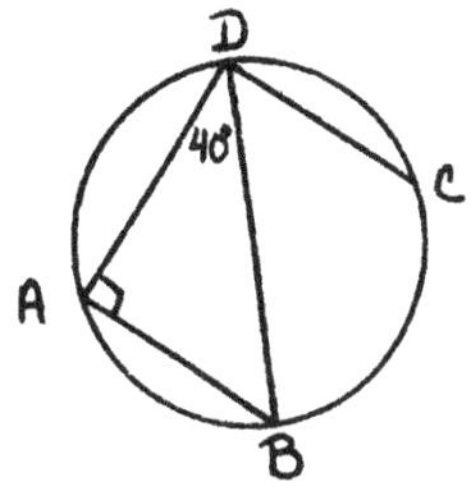

(Questions 19-20)

21. Given: circle B ≅ circle D

$\widehat{AE} \cong \widehat{CE}$

Prove: ABCD is a parallelogram.

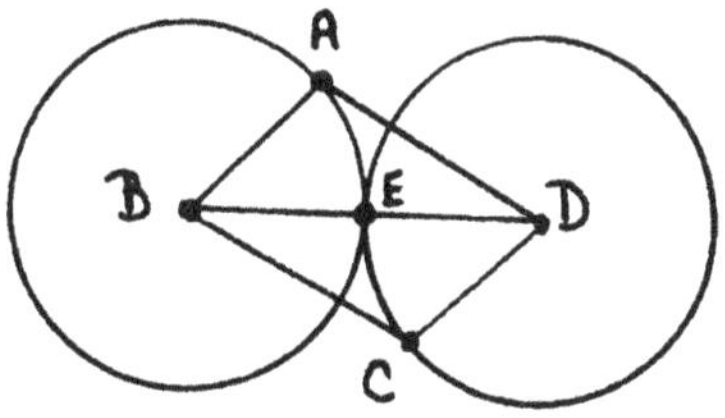

STATEMENTS	REASONS

Chapter 7 Test

Form 3

Decide if each of the following statements is true or is false. Then mark T or F in the space provided.

_________ 1. If the length of a chord of a circle is twice the length of the radius of the circle, then the chord is a diameter.

_________ 2. The measure of a central angle is equal to the measure of its intercepted arc.

_________ 3. Chords equidistant from the center of their circle are congruent.

_________ 4. The measure of an angle formed by the intersection of two secants outside a circle is equal to one-half the difference of the measures of the intercepted arcs.

_________ 5. The perpendicular bisector of the chord bisects the chords intercepted arc.

_________ 6. If a chord measures 10 inches, its intercepted arc has measure 10 inches.

_________ 7. Secants drawn to a circle from a point outside the circle are congruent.

_________ 8. If two circles are tangent, then their point of tangency and the center of each circle are collinear.

_________ 9. If two diameters intersect within a circle, their point of intersection is the center of the circle.

_________10. Opposite angles of an inscribed quadrilateral are congruent.

Solve each problem and place the answer in the space provided.

_________11. Find m($\angle$CEB).

_________12. Find m($\widehat{AB}$).

_________13. Find m($\widehat{AG}$).

_________14. Find m($\angle$EDC).

_________15. Find m($\overline{BD}$) if m($\overline{CD}$) = 10, m($\overline{DF}$) = 4, and m($\overline{DG}$) = 2.

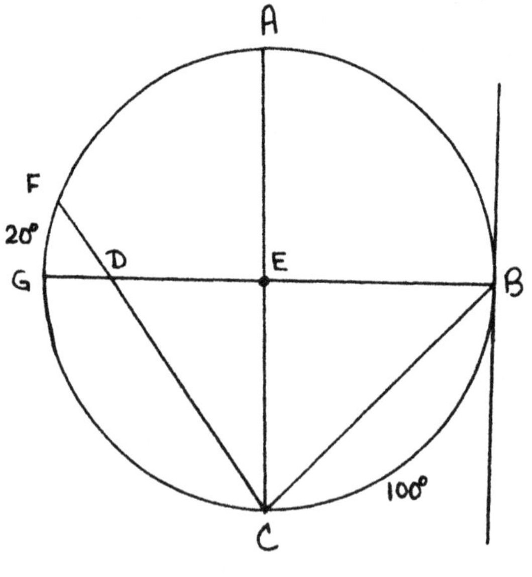

(Questions 11-15)

In 16 – 18 $m(\widehat{AB}) = 30°$, $m(\widehat{BC}) = 40°$, and $m(\widehat{CD}) = 50°$.

________16. Find $m(\angle E)$.

________17. Find $m(\angle F)$.

________18. Find $m(\angle G)$.

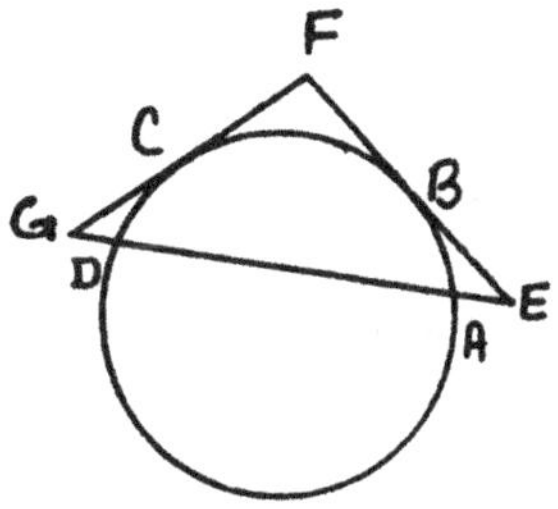

(Questions 16-18)

________19. $\overline{AB} \parallel \overline{DC}$. Find $m(\widehat{BC})$.

________20. $\overline{AB} \parallel \overline{DC}$. Find $m(\angle CDB)$.

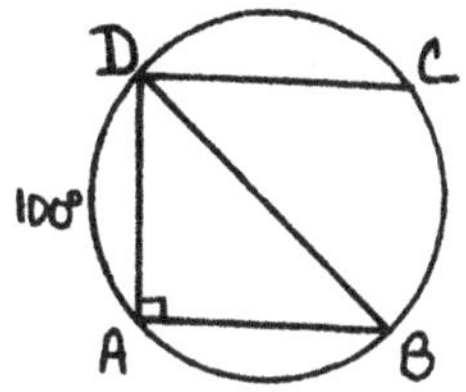

(Questions 19-20)

21. Given: Circle O

$\angle B \cong \angle C$

Prove: $\widehat{AB} \cong \widehat{AC}$

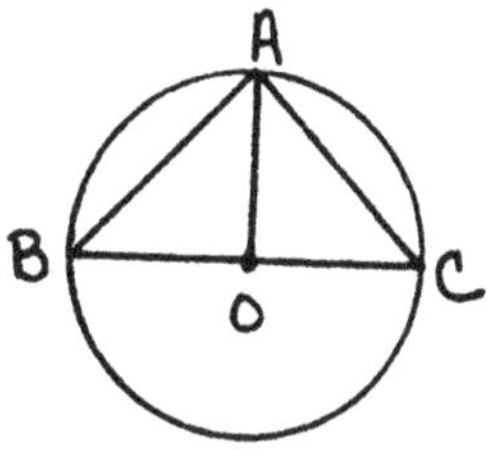

STATEMENTS	REASONS

CHAPTER 8

INEQUALITIES

SUPPLEMENTARY EXERCISES

1. Prove that any side of a triangle is less than one-half the triangle's perimeter.

2. Prove that the altitude upon the base of an isosceles triangle is less than one of the equal sides.

3. Given: Circle O with line $\overline{OB}$ passing through center O, tangent to $\overline{AB}$.

   Prove: $m(\overline{AB}) > m(\overline{BN})$.

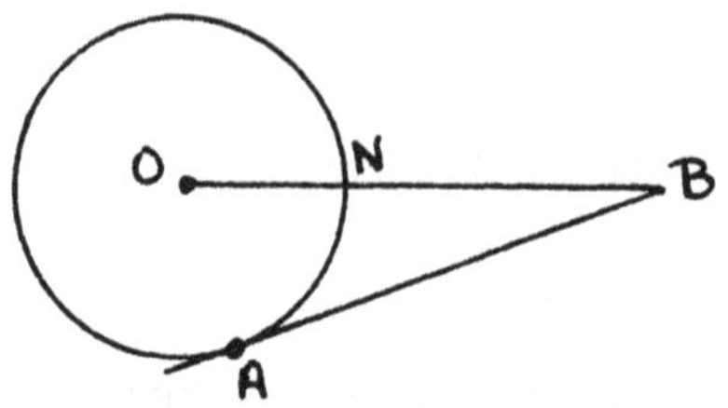

4. Given: $\triangle ABC$ with $m(\overline{BC}) < m(\overline{BA})$.
   $m(\overline{AE}) = m(\overline{CD})$.

   Prove: $m(\overline{CE}) < m(\overline{AD})$.

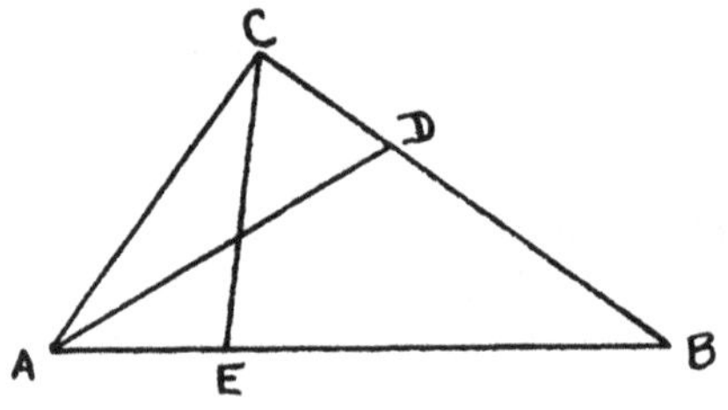

5. Given: Circle O with chords $\overline{AB}$ and $\overline{CD}$.
   $m(\overline{AB}) > m(\overline{CD})$.
   $\overline{MFON}$ is a diameter.

   Prove: $m(\angle 1) < m(\angle 2)$.
   (Hint: Consider the circle with diameter $\overline{OF}$.

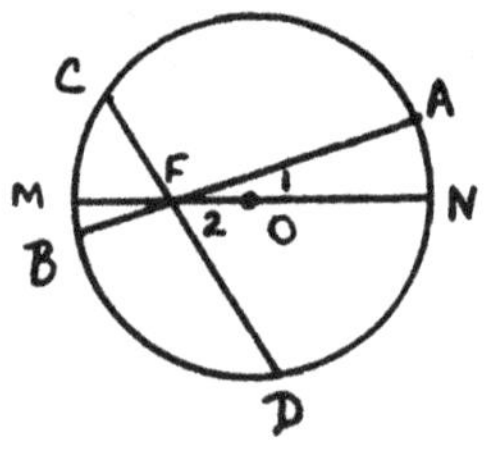

6. Given: $\triangle ABC$ with $m(\overline{AB}) < m(\overline{AC})$.
$\overrightarrow{BD}$ bisects $\angle ABC$.
$\overrightarrow{CD}$ bisects $\angle ACB$.

Prove: $m(\overline{BD}) < m(\overline{DC})$.

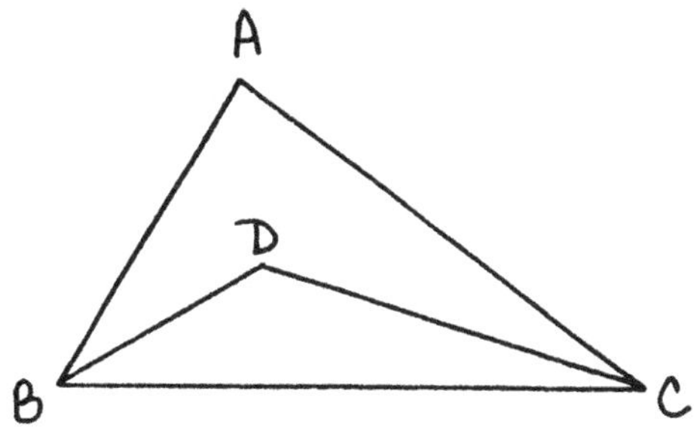

7. Given: $\triangle ABC$ is isosceles, with base $\overline{BC}$ and C is the midpoint of $\overline{BD}$.

Prove: $m(\overline{AD}) > m(\overline{AB})$.

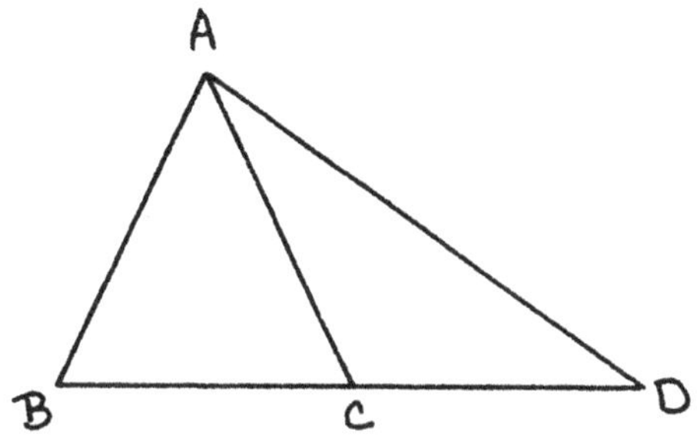

Chapter 8 Test

Form 1

Decide if each of the following statements is true or is false. Then mark T or F in the space provided.

_________ 1. If a and b are two numbers, then either $a > b$ or $b > a$.

_________ 2. If you add unequals to unequals, the sums will always be unequal.

_________ 3. The commutative law of addition states that for any numbers a and b, $a + b = b + a$.

_________ 4. The relation " $>$ " is a transitive relation.

_________ 5. In a triangle an exterior angle is greater than each interior angle.

_________ 6. The largest side of a triangle is always opposite the largest angle.

_________ 7. The greater of two chords in a circle is nearer the center of the circle.

_________ 8. The greater of two chords of a circle intercepts the greater arc.

_________ 9. If a right triangle is inscribed in a circle, the hypotenuse of the triangle is a diameter of the circle.

_________ 10. The larger of two central angles in a circle intercepts the greater arc.

Solve each problem and place the answer in the space provided.

_________ 11. Solve the inequality $2x - 5 < x + 4$.

_________ 12. Solve the inequality $3(x + 12) > 45 + 2x$.

_________ 13. Solve the inequality $\frac{x-3}{2} > x - 2$.

_________ 14. Solve the inequality $7x + 3 > \frac{3}{4}x + 28$.

_________ 15. Find the greatest angle of $\triangle$ ABC. (Do not be misled by the figure.)

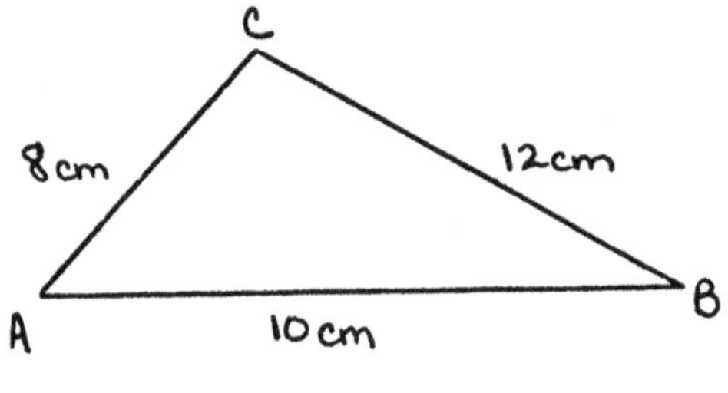

_________ 16. How large must a central angle be so that the chord intercepted by the central angle will equal a radius of the circle?

________17. Which chord is nearest the center of the circle?
m($\overline{AB}$) = 13 in., m($\overline{CD}$) = 12 in.,
and m($\overline{EF}$) = 10 in.
(Do not be misled by the figure.)

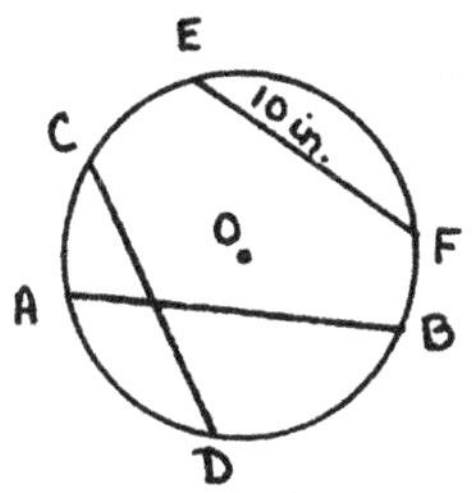

(Questions 17-18)

________18. Of the arcs intersected by the three chords, which is the greatest arc of circle O?
m($\overline{AB}$) = 13 in., m($\overline{CD}$) = 12 in.,
and m($\overline{EF}$) = 10 in.

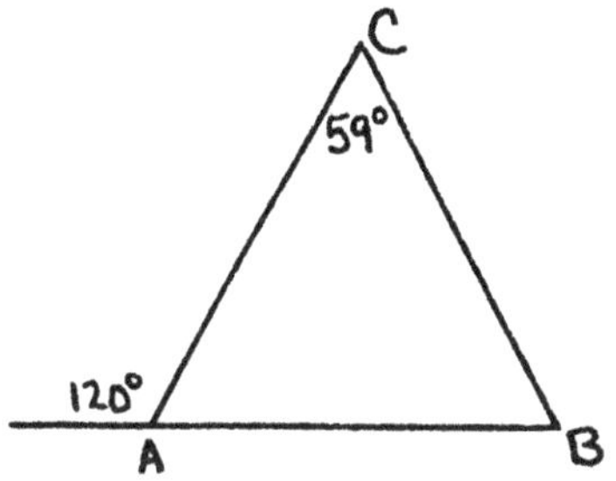

________19. In the figure, find the larger of $\overline{AC}$ and $\overline{BC}$.

________20. In any triangle, must the sum of any two sides be greater than the third side?

21. Given: Circle O.
m($\overline{AB}$) < m($\overline{CD}$).

Prove: m(∠1) < m(∠2).

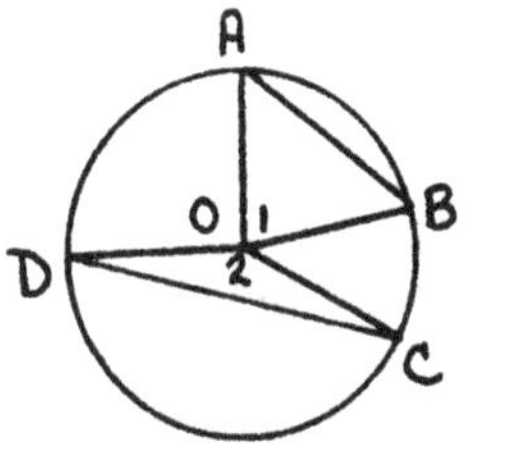

STATEMENTS	REASONS

Chapter 8 Test

Form 2

Decide if each of the following statements is true or is false. Then mark T or F in the space provided.

_________ 1. If a is not less than b, and a is not equal to b, then a is greater than b.

_________ 2. The associative law of multiplication states that for any numbers a, b, and c, $a(b+c) = ab + ac$.

_________ 3. If $a > b$ and $b > c$ then $c > a$.

_________ 4. If $a < b$ and $c < d$, then $a + c < b + d$.

_________ 5. If $a < b$ and $c < d$, then $ac < bd$.

_________ 6. The angle with the largest measure is always opposite the side with the largest measure.

_________ 7. The measure of the hypotenuse of a right triangle is greater than the sum of the measures of the legs.

_________ 8. In congruent circles, the greater of two minor arcs has the shorter chord.

_________ 9. In the same circle, the shorter of two chords is the greatest distance from the center of the circle.

_________ 10. If a central angle of a circle has a measure of 90°, then the chord intercepted by the angle is longer than a radius of the circle.

Solve each problem and place the answer in the space provided.

_________ 11. Solve the inequality $3x - 7 < 2x + 3$.

_________ 12. Solve the inequality $5(x-6) > 3x + 2$.

_________ 13. Solve the inequality $\frac{2x+1}{3} < x - 4$.

_________ 14. Solve the inequality $\frac{1}{2}x + 4 < x - 3$.

_________ 15. Find the smallest angle of $\triangle$ ABC. (Do not be misled by the figure.)

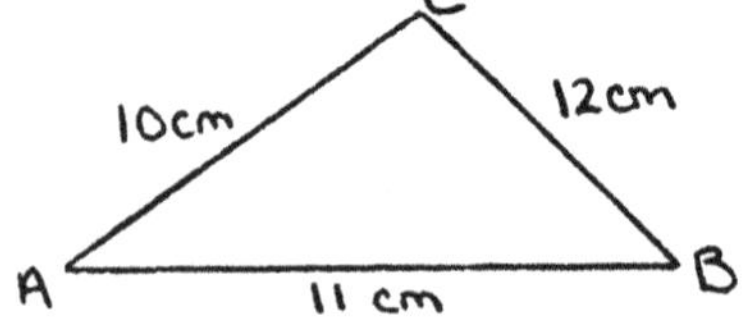

_________ 16. If m($\overparen{AB}$) = 80° and m($\overparen{CD}$) = 40°, which of $\angle 1$ and $\angle 3$ is larger?

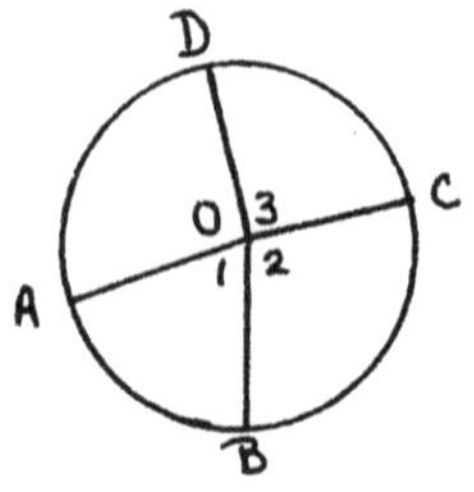

_________17. Which chord is nearest the center of the circle?
$m(\overline{AB}) = 10$ in., $m(\overline{CD}) = 15$ in.,
and $m(\overline{EF}) = 11$ in.
(Do not be misled by the figure.)

_________18. Of the arcs intersected by the three chords, which is the smallest arc of circle O?
$m(\overline{AB}) = 10$ in., $m(\overline{CD}) = 15$ in.,
and $m(\overline{EF}) = 11$ in.

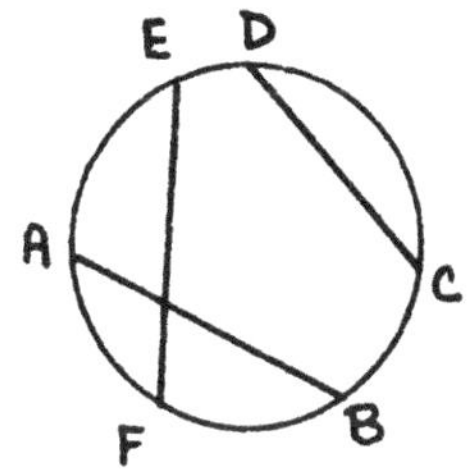

(Questions 17-18)

_________19. In the figure, find the largest side of $\triangle ABC$.

_________20. In the figure, find the smallest side of $\triangle ABC$.

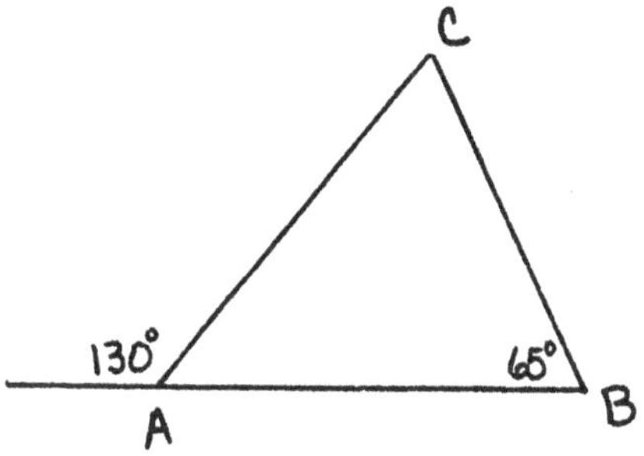

21. Given: $\overline{BD}$ is a median.
$m(\overline{AD}) > m(\overline{CD})$.

Prove: $m(\angle 1) > m(\angle 2)$.

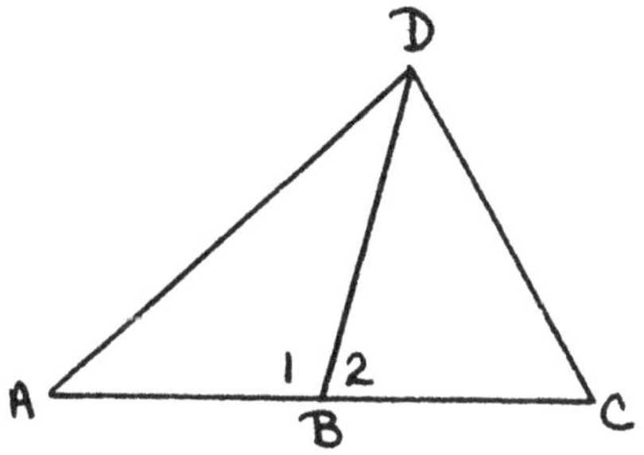

STATEMENTS	REASONS

Chapter 8 Test

Form 3

Decide if each of the following statements is true or is false. Then mark T or F in the space provided.

________ 1. If $a \neq b$, then $a > b$.

________ 2. The associative law of addition states that for any numbers a, b, and c, $a + (b + c) = (a + b) + c$.

________ 3. If unequal quantities are multiplied to unequal quantities, then the products are unequal quantities.

________ 4. If $a < b$ and $b < c$, then $a < c$.

________ 5. If the base angles of an isosceles triangle are greater than 60°, the base is the shortest side of the triangle.

________ 6. The longer of two chords in a circle is nearer the center of the circle.

________ 7. A chord can be greater than a diameter of the circle.

________ 8. The larger of two inscribed angles in a circle intercepts the greater arc.

________ 9. If a central angle of a circle has a measure of 45°, then the chord intercepted by the angle is longer than the radius of the circle.

________10. If $m(\overset{\frown}{DC}) > m(\overset{\frown}{AB})$, then $\overline{AB}$ is nearer to the circle.

Solve each problem and place the answer in the space provided.

________11. Solve the inequality $5x - 7 > x + 9$.

________12. Solve the inequality $2x + 3 < 3(x - 9)$.

________13. Solve the inequality $\frac{x+6}{2} > x - 8$.

________14. Solve the inequality $\frac{2}{3}x + 1 < x - 6$.

________15. Find the largest angle of $\triangle$ ABC. (Do not be misled by the figure.)

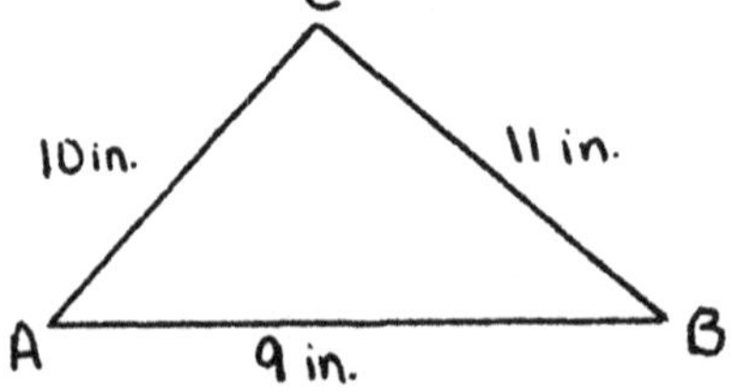

________16. If $m(\overline{AB}) = 9$ cm, $m(\overline{CD}) = 12$ cm, and $m(\overline{EF}) = 15$ cm, which chord is nearest to the center of the circle. (Do not be misled by the figure.)

________17. If $m(\overline{AB}) = 9$ cm, $m(\overline{CD}) = 12$ cm, and $m(\overline{EF}) = 15$ cm, which chord cuts the greatest minor arc? (Do not be misled by the figure.)

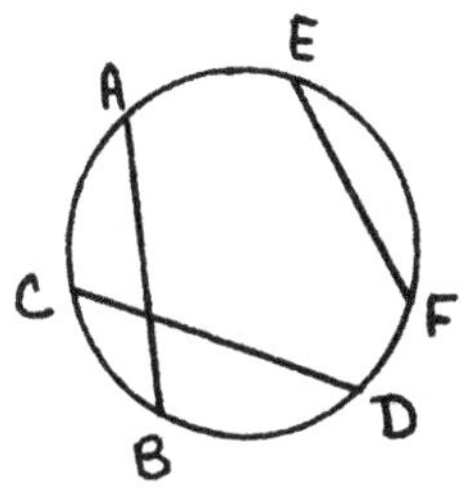

(Questions 16-17)

________18. In the figure, find the smallest side of $\triangle ABC$.

________19. In the figure, find the largest side of $\triangle ABC$.

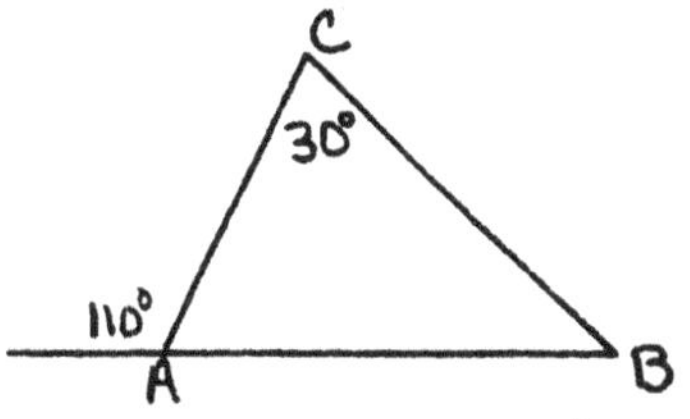

(Questions 18-19)

________20. Does a triangle with sides 5 cm, 4 cm, and 10 cm exist?

21. Given: $\overline{AB} \cong \overline{AD}$.

$m(\angle BAC) > m(\angle DAC)$.

Prove: $m(\overline{BC}) > m(\overline{DC})$.

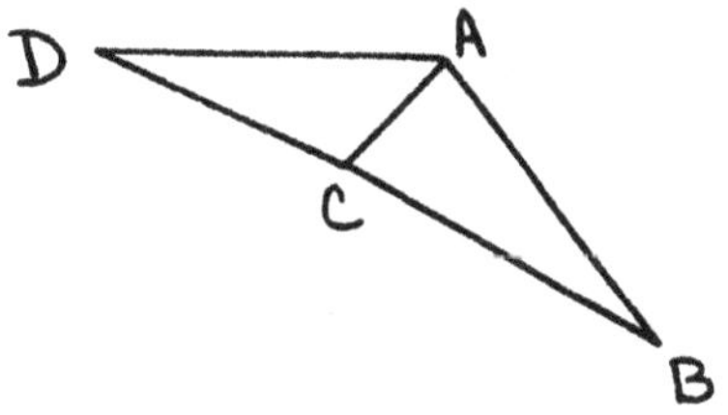

STATEMENTS	REASONS

CHAPTER 9

GEOMETRIC LOCI

SUPPLEMENTARY EXERCISES

1. An equilateral triangle is inscribed in a circle of radius a. What is the length of a side? What is the area?

$$\left(a\sqrt{3} \ , \ \frac{3\sqrt{3}a^2}{4} \right)$$

2. Two circles are tangent externally. Find the locus of points P for which the tangents from P to the circles are equal.

(The locus of points P is the line tangent to the two circles at their point of tangency.)

3. Find the locus of the center of a quarter that rolls around the edge of a penny.

(The locus is a circle with the same center as the penny and with radius equal to the sum of the radii of the two coins.)

4. Sketch the locus of points which are equidistant from a point F and from a given line not containing F.

5. Sketch the locus of the midpoints of all chords passing through a point P lying on a given circle. Does the locus contain the point P? (no)

6. Describe the locus of the midpoint of a chord of given length a inscribed in a circle of radius r. Discuss the situation when $a < 2r$ and when $a = 2r$. Can $a > 2r$?

7. Construct the locus of the vertex angle of a triangle on a fixed base if the vertex angle is always equal to

(a) 135°

(b) 120°

8. Given $\triangle$ ABC with centroid F.

$m(\overline{FE}) = 6$ cm, $m(\overline{DE}) = 9$ cm, $m(\overline{CE}) = 18$ cm

Find: (a) $m(\overline{AD})$ ($9\sqrt{3}$ cm)

(b) the area of $\triangle$ ABC ($162\sqrt{3}$ sq cm)

(c) the area of $\triangle$ ACE ($81\sqrt{3}$ sq cm)

(d) $m(\angle DAE)$ (30°)

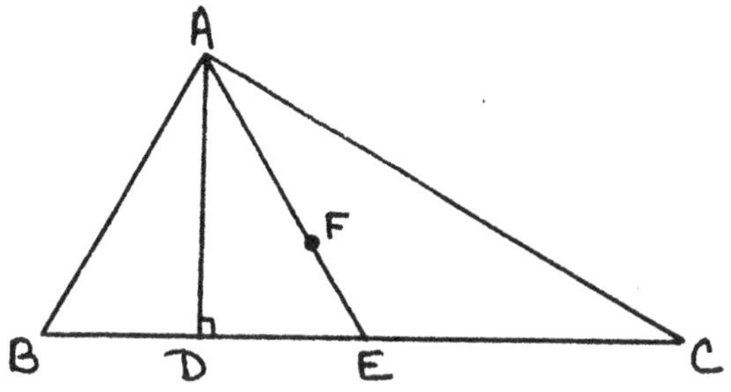

Chapter 9 Test

Form 1

Decide if each of the following statements is true or is false. Then mark T or F in the space provided.

________ 1. The locus of points equidistant from a fixed point and in a single plane is a circle.

________ 2. The locus of points equidistant from two fixed points and in a single plane is any bisector of the segment joining the two fixed points.

________ 3. An angle bisector is the locus of all points equidistant from the sides of an angle.

________ 4. All bisectors of the three sides of a triangle meet in a point.

________ 5. The altitudes of a triangle always meet at a point inside the triangle.

________ 6. The medians of a triangle always meet at a point inside the triangle.

________ 7. The point of intersection of the angle bisectors of the angles of a triangle is called the incenter of the triangle.

________ 8. The point where the medians of a triangle intersect is called the centroid of the triangle.

________ 9. A circle can be constructed through any four fixed points.

________10. A square can be inscribed within a circle with a compass and a straightedge.

________11. The point where the altitudes of a triangle meet is called the orthocenter of the triangle.

________12. The point where the angle bisectors of the angles of a triangle meet is called the orthocenter of the triangle.

________13. The point of intersection of the perpendicular bisectors of the sides of a triangle is called the circumcenter.

________14. To determine a locus, both a statement and its converse must be proved.

________15. The locus of the vertex of the right angle of a right triangle with a fixed hypotenuse falls on a arc of a circle.

Solve each problem and place the answer in the space provided.

________16. Find $m(\overline{GD})$ if $\overline{BF}$, $\overline{AE}$, and $\overline{CD}$ are medians and $m(\overline{CG}) = 6$ cm.

________17. Find $m(\overline{AG})$ if $\overline{BF}$, $\overline{AE}$, and $\overline{DC}$ are medians and $m(\overline{GE}) = 2$ cm.

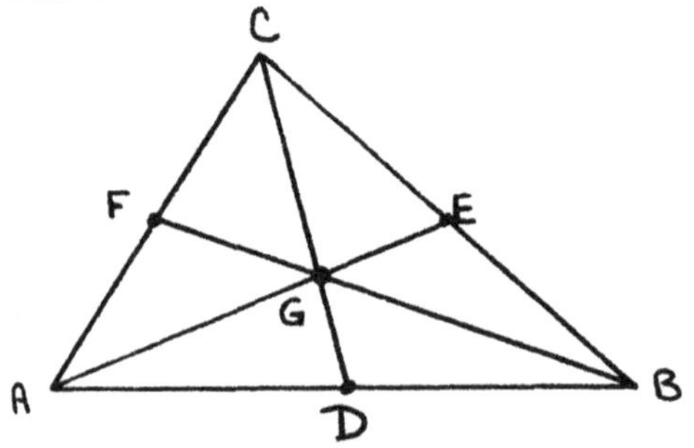

(Questions 16–17)

_________18. If the angle bisectors of an equilateral triangle are 9 in. long, find the radius of the circumscribed circle.

_________19. What is the locus of points equidistant from the vertices of a triangle?

_________20. Determine if a triangle with sides of 24 in., 30 in., and 18 in., is a right triangle.

_________21. Using a compass and a straightedge, construct the locus of points $\frac{1}{2}$ in. from segment $\overline{AB}$.

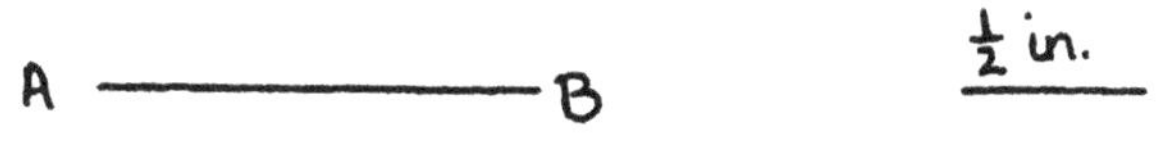

_________22. Construct the locus of points $\frac{1}{2}$ in. from point P and equidistant from the sides of $\angle ABC$.

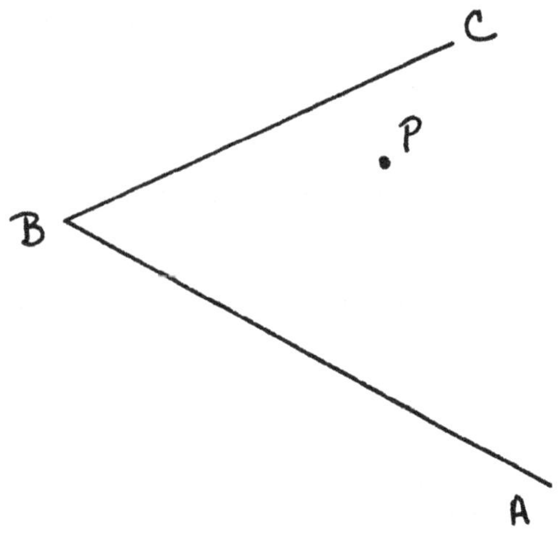

Chapter 9 Test

Form 2

Decide if each of the following statements is true or is false. Then mark T or F in the space provided.

________ 1. The locus of points that are one inch from a given point and in a single plane is a circle with diameter of 2 inches.

________ 2. The locus of points less than 5 cm from a given point A is the interior of a circle with its center at A and a radius of 5 cm.

________ 3. The locus of all points equidistant from 2 points is any bisector of the segment joining the 2 points.

________ 4. The locus of points a given distance from a given line is a pair of lines, one on each side of the given line, parallel to the given line and at the given distance from it.

________ 5. The point of intersection of the altitudes of a triangle is called the circumcenter of the triangle.

________ 6. A circumscribed circle is a circle inside a triangle which is tangent to the three sides of the triangle.

________ 7. The medians of a triangle always meet at a point inside the triangle.

________ 8. The point of intersection of the medians of a triangle is two-thirds of the way from the vertex to the midpoint of the opposite side.

________ 9. It is always possible to construct a circle passing through three points in a plane.

________ 10. The point of intersection of the medians of a triangle is called the centroid of the triangle.

________ 11. The point of intersection of the altitudes of a triangle is called the orthocenter of the triangle.

________ 12. The perpendicular bisectors of the sides of a triangle meet at a point that is equally distant from the sides of the triangle.

________ 13. The altitudes of a triangle may meet at a point outside the triangle.

________ 14. The locus of a point at a given distance from a given line is either the line parallel to the given line that is on one side, or the other side of the given line.

________ 15. The locus of a point equidistant from the sides of an angle is the bisector of the angle.

Solve each problem and place the answer in the space provided.

________ 16. Find $m(\overline{AG})$ if $\overline{CF}$, $\overline{AD}$, and $\overline{BE}$ are medians and $m(\overline{AD}) = 9$ m, $m(\overline{GE}) = 5$ m, and $m(\overline{GC}) = 8$ m.

________ 17. Find $m(\overline{BE})$ if $\overline{CF}$, $\overline{AD}$, and $\overline{BE}$ are medians. and $m(\overline{AD}) = 9$ m, $m(\overline{GE}) = 5$ m, and $m(\overline{GC}) = 8$ m.

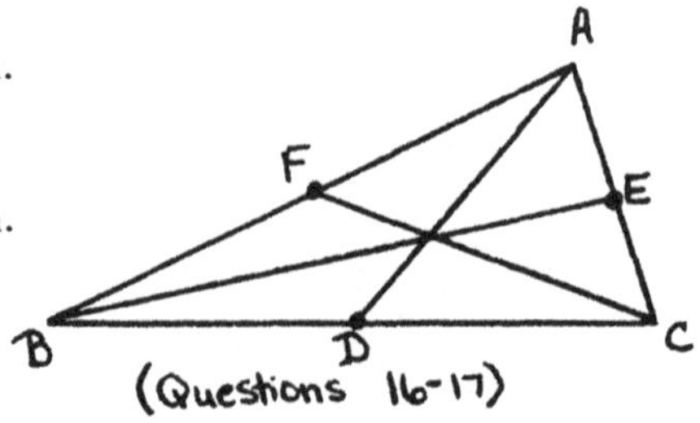

(Questions 16-17)

________18. If an altitude of an equilateral triangle is 12 in. long, find the radius of the circumscribed circle.

________19. If an altitude of an equilateral triangle is 12 in long, find the radius of the inscribed circle.

________20. Determine the locus of points equidistant from 3 noncolinear points.

21. Using a compass and a straightedge, construct the circle through the given three noncollinear points.

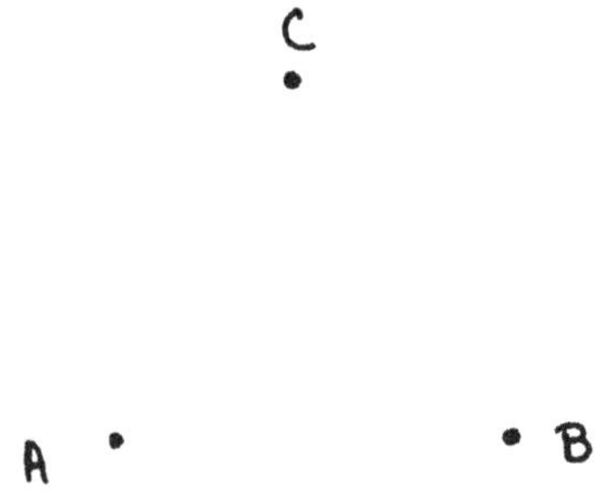

22. Construct the locus of points 1 in. from point P and equidistant from the sides of $\angle$ABC.

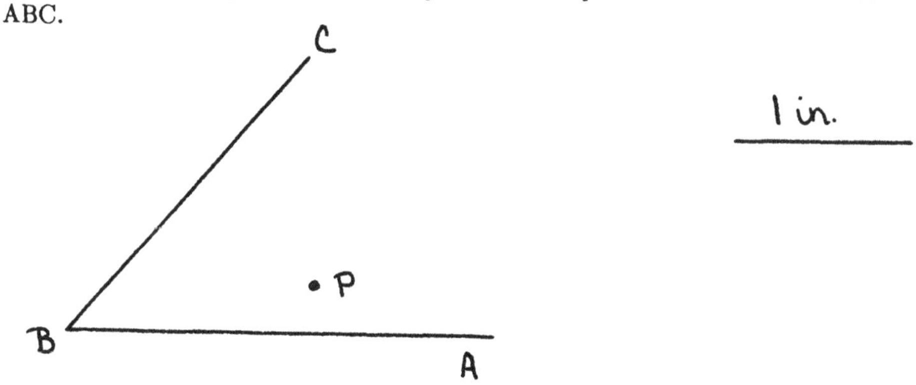

Chapter 9 Test

Form 3

Decide if each of the following statements is true or is false. Then mark T or F in the space provided.

_________ 1. In a single plane, the locus of points that are one inch from a circle with a radius of 5 in. is a circle whose center is the same as the given circle and whose radius is 4 in.

_________ 2. In a single plane, the locus of points equidistant from two parallel lines is a line parallel to the given lines and is midway between the given lines.

_________ 3. The locus of all points equidistant from two points is the perpendicular bisector of the segment joining the two points.

_________ 4. In an equilateral triangle the centroid, orthocenter, incenter, and circumcenter coincide at the same point.

_________ 5. The altitudes of a triangle always meet at a point inside the triangle.

_________ 6. A circumcenter is the center of a circle that passes through the three vertices of the triangle.

_________ 7. The point of intersection of the three angle bisectors of a triangle is called the incenter.

_________ 8. The orthocenter is the point you construct to find the center of the circle passing through three noncolinear points.

_________ 9. The incenter and the circumcenter of a triangle can never be the same point.

_________10. The center of a circle tangent to the three sides of a triangle is called the incenter.

_________11. The circumcenter is a point that is always inside the triangle.

_________12. The perpendicular bisectors of the sides of a triangle meet at a point that is equally distant from the vertices of the triangle.

_________13. The locus of the vertex of the right angle of a right triangle with a fixed hypotenuse is a circle with the hypotenuse as its diameter, minus the endpoints of the diameter.

_________14. The locus of points equidistant from the four vertices of a square is one point.

_________15. Given segment $\overline{AB}$, the locus of points that are the vertices of isosceles triangle having $\overline{AB}$ as a base is the perpendicular bisector of $\overline{AB}$ minus the midpoint of $\overline{AB}$.

Solve each problem and place the answer in the space provided.

_________16. Find $m(\overline{AG})$ if $\overline{CF}$, $\overline{AD}$, and $\overline{BE}$ are medians and $m(\overline{AD}) = 12$ m, $m(\overline{GE}) = 6$ m, and $m(\overline{GC}) = 9$ m.

_________17. Find $m(\overline{BE})$ if $\overline{CF}$, $\overline{AD}$, and $\overline{BE}$ are medians. and $m(\overline{AD}) = 12$ m, $m(\overline{GE}) = 6$ m, and $m(\overline{GC}) = 9$ m.

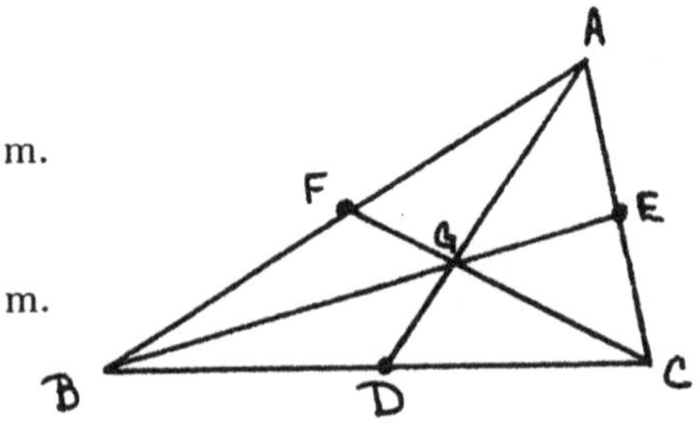

(Questions 16-17)

________18. If the angle bisectors of an equilateral triangle are 15 cm long, find the radius of the circumscribed circle.

________19. If the perpendicular bisectors of the sides of an equilateral triangle are 9 m long, find the radius of the inscribed circle.

________20. Determine the locus of the center of a circle of radius 2 in that rolls around a circle with radius 5 in.

21. Using a compass and a straightedge, construct the inscribed circle of the given triangle.

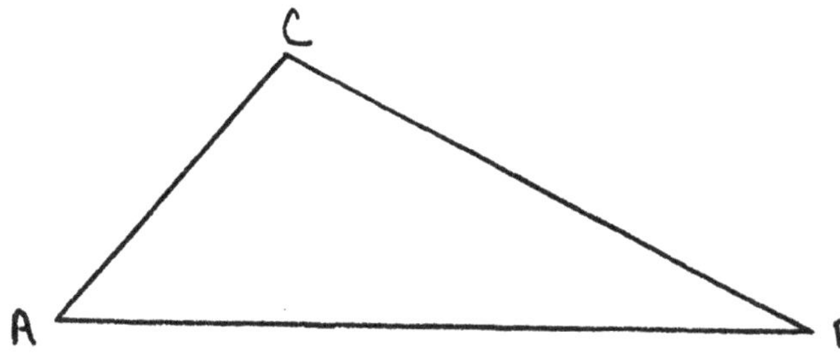

22. Sketch the locus of the midpoints of all chords of a circle that have a fixed point of the circle as an endpoint.

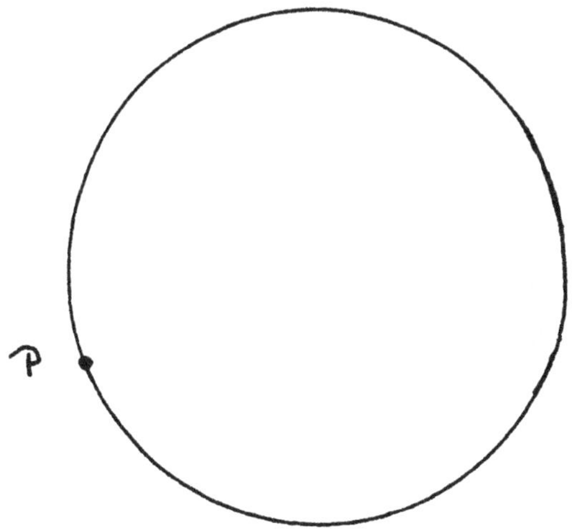

CHAPTER 10

REGULAR POLYGONS AND CIRCLES

SUPPLEMENTARY EXERCISES

1. Construct a regular hexagon and inscribe a circle within it.

2. If the radius of a circle inscribed in a regular hexagon is a, what is the radius of a circumscribed circle?

$\left(\frac{2\sqrt{3}a}{3}\right)$

3. How many circles of radius 5, tangent to each other, may be tangent to another circle of radius 5? (6)

4. Prove that the measure of the apothem of an equilateral triangle is one-third the measure of the altitude of the triangle.

5. Given: Hexagon ABCDEF inscribed in circle O.

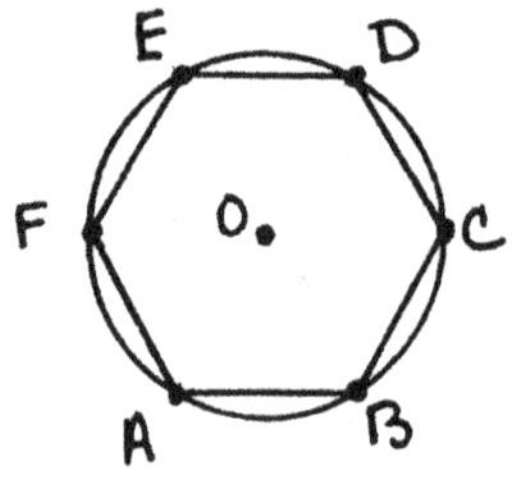

Prove: $m(\angle A) + m(\angle C) + m(\angle E) = 360°$.

6. A regular hexagon has sides of length 10 ft. What is the area of an equilateral triangle formed by joining every other vertex? ($75\sqrt{3}$ sq ft)

7. An equilateral triangle is inscribed in a circle of radius 2. What is the area of the triangle? ($3\sqrt{3}$ sq units)

8. A circle of radius 1 is inscribed in an equilateral triangle. Circumscribed about this triangle is another circle. What is the area between the circles? (3π sq units)

Chapter 10 Test

Form 1

Decide if each of the following statements is true or is false. Then mark T or F in the space provided.

_________ 1. An regular n-sided polygon inscribed in a circle divides the circle into n equal arcs.

_________ 2. A circle can be circumscribed about any regular polygon.

_________ 3. An apothem of a regular polygon is a line segment drawn from the center of the polygon perpendicular to one of the polygon's sides.

_________ 4. The circumference of a circle is given by the formula $C = \pi r^2$.

_________ 5. If a regular hexagon (six-sided figure) has sides of 6 cm and its apothem is $3\sqrt{3}$ cm, the area of the hexagon is $18\sqrt{3}$ sq cm.

_________ 6. If a square has sides of 5 in. and an apothem of a in., its area is $10a$ in.

_________ 7. The area of a circle with a circumference of 14 in. is 14π sq in.

_________ 8. If a circle has a radius of 2 ft, its circumference is numerically equal to its area.

_________ 9. A circle can be inscribed in a rectangle that is not a square.

_________10. A regular polygon with a central angle of 45° can be constructed.

Solve each problem and place the answer in the space provided.

_________11. Find the area of a circle whose circumference is 16π.

_________12. Find the number of sides of an inscribed regular polygon if its central angle is 60°.

_________13. Find the area of the regular polygon.

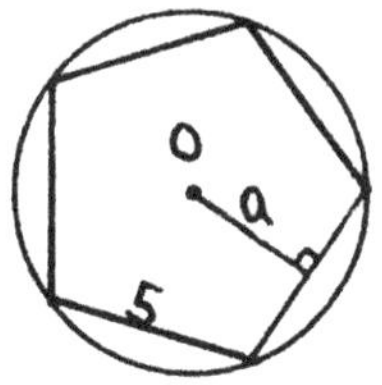

_________14. Find the measure of a central angle of an inscribed regular octagon (8-sided figure).

_________15. Find the radius of a circle that has an inscribed regular hexagon within it with a perimeter of 30 cm.

_________16. Find the area of a regular decagon (10-sided figure) with side 8 in. long and an apothem 12 in. long.

_________17. Is a rhombus a regular polygon?

________18. Is a rectangle a regular polygon?

________19. Find the radius of a circle if its area is 81π sq in.

________20. Find the measure of the central angle of a regular decagon (10 sided).

21. With a compass and a straightedge, draw a circle and inscribe a regular octagon in it.

22. Draw a triangle and circumscribe a circle about it.

Chapter 10 Test

Form 2

Decide if each of the following statements is true or is false. Then mark T or F in the space provided.

_________ 1. An n-sided polygon inscribed in a circle divides the circle into n equal areas.

_________ 2. A circle can be inscribed in any regular polygon.

_________ 3. The radius of a regular polygon is a line segment drawn from the center of the polygon to one of its vertices.

_________ 4. The area of a circle is given by the formula $A = 2\pi r$.

_________ 5. The center of a circumscribed regular polygon coincides with the center of its inscribed circle.

_________ 6. The apothem of a regular polygon is the radius of the inscribed circle of the polygon.

_________ 7. The central angle of a regular octagon has measure of 45°.

_________ 8. The circumference of a circle with diameter of 4 in. long is 8π in.

_________ 9. The area of a circle with radius of $\frac{1}{2}$ cm long is numerically equal to its circumference.

_________ 10. A circle can be inscribed in a rectangle that is not a square.

Solve each problem and place the answer in the space provided.

_________ 11. Find the circumference of a circle whose area is 25π sq m.

_________ 12. Find the number of sides of an inscribed regular polygon if its central angle is 36°.

_________ 13. Find the area of the regular polygon.

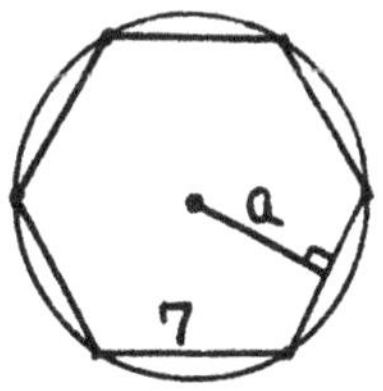

_________ 14. Find the measure of a central angle of an inscribed regular pentagon (5-sided figure).

_________ 15. Find the radius of a circle that has an inscribed regular hexagon within it with a perimeter of 42 m.

_________ 16. Find the area of a regular hexagon (6-sided figure) with side 8 in. long and an apothem $4\sqrt{3}$ in. long.

_________ 17. Is an isosceles triangle a regular polygon?

_________ 18. A circle has a radius that is 4 in. long. If a sector of the circle has an arc that measures 30°, find the area of the sector.

________19. Find the radius of a circle if its area is 49π sq m.

________20. A rectangle has dimensions of π cm by 16 cm. Find the radius of a circle with the same area.

21. With a compass and a straightedge, draw a circle and inscribe a regular hexagon in it.

22. Construct a square and circumscribe a circle about it.

Chapter 10 Test

Form 3

Decide if each of the following statements is true or is false. Then mark T or F in the space provided.

_________ 1. A circle can be circumscribed about any regular polygon.

_________ 2. A rhombus that is not a square can be inscribed in a circle.

_________ 3. The apothem of a regular polygon bisects its respective side.

_________ 4. The radius of a regular polygon is also the radius of the circle circumscribed about the polygon.

_________ 5. The circumference of a circle can be found by the formula $C = d\pi$. Where d is the diameter of the circle.

_________ 6. A sector of a circle is a triangle.

_________ 7. The central angle of a pentagon is 72°.

_________ 8. The circumference of a circle with a radius that is 6 cm long is 12π cm.

_________ 9. The circumference of a circle with radius 1 in. is numerically equal to its area.

_________ 10. In congruent circles, sectors that have central angles with the same measure have equal areas.

Solve each problem and place the answer in the space provided.

_________ 11. Find the area of a circle whose circumference is 10π cm.

_________ 12. Find the perimeter of a regular polygon whose area is 50 sq in. and whose apothem is 4 in.

_________ 13. Find the area of the regular polygon.

_________ 14. Find the measure of a central angle of an inscribed regular decagon (10-sided figure).

_________ 15. Find the radius of a circle that has an inscribed regular hexagon within it with a perimeter of 48 cm.

_________ 16. Find the area of a regular hexagon (6-sided figure) with side 10 in. long and an apothem $5\sqrt{3}$ in. long.

_________ 17. Is an parallelogram a regular polygon?

________18. A circle has a radius that is 8 cm long. If a sector of the circle has an arc that measures 90°, find the area of the sector.

________19. Find the diameter of a circle if its area is 64π sq m.

________20. A rectangle has dimensions of π ft by 25 ft. Find the radius of a circle with the same area.

21. With a compass and a straightedge, construct an equilateral triangle and circumscribe a circle about it.

22. Construct a regular octagon and inscribe a circle within it.

CHAPTER 11

MATHEMATICAL LOGIC

SUPPLEMENTARY EXERCISES

Find the strongest possible conclusion. If none is possible, write no conclusion.

1. None but the brave deserve to be called heros.
   Mr. X won an award for bravery for rescuing a child from danger.

2. If the nail works loose from the wall, then the barometer falls.
   If the barometer falls, then bad weather is ahead.
   (Hint: How is the word "falls" being used?)

3. If I plant the seeds in May, then if the seed catalog is accurate, I'll have flowers in June.
   I planted the flowers in May.
   I had no flowers in June.

4. If the road is not plowed, I will not be able to get to work.
   I can't get to work.

5. My saucepans are the only things I have that are made of tin;
   I find all your presents very useful;
   None of my saucepans are of the slightest use.

6. No ducks waltz;
   No officers ever decline to waltz;
   All my poultry are ducks.

7. The only food that my doctor allows me is food that is not very rich;
   Nothing that agrees with me is unsuitable for supper;
   Wedding cake is always very rich;
   My doctor allows me all food that is suitable for supper.

Problem 5 – 7 are attributed to Lewis Carroll.

Chapter 11 Test

Form 1

Decide if each of the following statements is true or is false. Then mark T or F in the space provided.

________ 1. The argument known as the law of detachment has the form [$p \Rightarrow q$ and p] therefore q.

________ 2. If p is true and q is true, then $\sim p \vee q$ is true.

________ 3. A statement $p \Rightarrow q$ is logically equivalent to the statement $\sim p \Rightarrow \sim q$.

________ 4. A statement and its contrapositive are logically equivalent.

________ 5. The converse of $p \Rightarrow q$ is $q \Rightarrow p$.

________ 6. The inverse of $p \Rightarrow q$ is $\sim p \Rightarrow \sim q$.

________ 7. The contrapositive of $q \Rightarrow p$ is $\sim p \Rightarrow \sim q$.

________ 8. If we are given that "if it rains, then grass will grow" and we observe "the grass does grow", we can conclude that "it does rain".

________ 9. The clause following the word if in an "If . . , then . . " sentence is called the *given* or *antecedent.*

________10. The contrapositive of the contrapositive of a statement is the original statement.

Write each symbolic expression as an English sentence. Assume that the variables represent the following statements.

p: Laurie is friendly.
q: Marcia is happy.
r: John is grouchy.

11. $\sim p \Rightarrow \sim q$

____

12. $p \wedge (q \vee r)$

____

13. Write the contrapositive of "If a triangle has two congruent angles, then it has two congruent sides".

____

14. Construct a truth table to verify $\sim p \vee q \Leftrightarrow \sim (p \wedge \sim q)$ is a tautology.

In 15 – 17 determine the valid conclusion if one is warranted. If no valid conclusion is possible, write "no conclusion".

15. If it rains, then I will work inside.
I will work inside.

16. Either Joe is sick, or he is playing hooky.
Joe is not sick.

17. If I work hard, then I will succeed.
I do not succeed hard.

In 18 – 20 use Euler circles to determine, if possible, a valid conclusion to the given premises.

18. All logic students are mathematics students.
No mathematics students are gullible.

19. No 14 year olds can drive.
My friend can not drive.

20. All rectangles are parallelograms.
No triangles are parallelograms.

Chapter 11 Test

Form 2

Decide if each of the following statements is true or is false. Then mark T or F in the space provided.

_________ 1. Give me a break! is a statement.

_________ 2. If p is false and q is true, then $p \vee q$ is true.

_________ 3. If p is false and q is true, then $p \Leftrightarrow q$ is true.

_________ 4. If p is true and q is false, then $\sim p \wedge q$ is true.

_________ 5. The contrapositive of $\sim p \Rightarrow q$ is $\sim q \Rightarrow p$.

_________ 6. The inverse of $\sim p \Rightarrow \sim q$ is $\sim q \Rightarrow \sim p$.

_________ 7. $\sim (p \wedge q) \Leftrightarrow \sim p \vee \sim q$.

_________ 8. The negation of the statement "The puppy is cuddly and warm." is "The puppy is not cuddly or the puppy is not warm".

_________ 9. In a valid argument, if the premises is true, then the conclusion is true.

_________ 10. $p \Rightarrow q$ is a valid form of an argument.

$$\begin{array}{l} p \Rightarrow q \\ \underline{q \quad} \\ p \end{array}$$

Write each symbolic expression as an English sentence. Assume that the variables represent the following statements.

p: Laurie is friendly.
q: Marcia is happy.
r: John is grouchy.

11. $\sim p \wedge \sim q$

____

12. $p \wedge q \vee \sim r$

____

13. Write the contrapositive of "If quadrilateral is a square, then it is a rectangle."

____

14. Construct a truth table to verify $\sim (p \wedge q) \Leftrightarrow \sim p \vee \sim q$ is a tautology.

In 15 – 17 determine the valid conclusion if one is warranted. If no valid conclusion is possible, write "no conclusion".

15. If it rains, then I will work inside.
I do not work inside.

____

16. Either Joe is sick, or he is playing hooky.
Joe is not playing hooky.

____

17. If I work hard, then I will succeed.
I do not word hard.

____

In 18 – 20 use Euler circles to determine, if possible, a valid conclusion to the given premises.

18. All cars have four wheels.
No bikes have four wheels.

19. All instructors are understanding.
All understanding people are helpful.

20. All winners are happy.
No gamblers are winners.

Chapter 11 Test

Form 3

Decide if each of the following statements is true or is false. Then mark T or F in the space provided.

_________ 1. Joan is helpful. is a statement.

_________ 2. If p is true and q is false, then $p \wedge q$ is true.

_________ 3. If p is false and q is false, then $p \Rightarrow q$ is false.

_________ 4. If p is true and q is true, then $\sim (p \vee q)$ is false.

_________ 5. The inverse of $\sim p \Rightarrow q$ is $q \Rightarrow \sim p$.

_________ 6. The converse of $p \Rightarrow \sim q$ is $\sim q \Rightarrow p$.

_________ 7. $p \vee (p \wedge r) \Leftrightarrow (p \vee q) \wedge (p \vee r)$.

_________ 8. The negation of the statement "Either I stop at the store or I will come striaght home." is "I do not stop at the store and I do not come straight home".

_________ 9. In a valid argument, if the premises is false, then the conclusion must be false.

_________10. $\begin{array}{l} p \vee q \\ \underline{\sim q} \\ p \end{array}$ is a valid form of an argument.

Write each symbolic expression as an English sentence. Assume that the variables represent the following statements.

p: Laurie is friendly.
q: Marcia is happy.
r: John is grouchy.

11. $p \wedge \sim q$

____

12. $p \wedge \sim q \vee r$

____

13. Write the contrapositive of "If two triangles are congruent, then they have the same area."

____

14. Construct a truth table to verify $\sim (p \vee q) \Leftrightarrow \sim p \wedge \sim q$ is a tautology.

In 15 – 17 determine the valid conclusion if one is warranted. If no valid conclusion is possible, write "no conclusion".

15. If I pass the test, then I will pass the class.
I passed the class.

____

16. It will be sunny or rainy today.
It isn't rainy.

____

17. You can go in if and only if you have an invitation.
You have an invitation.

____

In 18 – 20 use Euler circles to determine, if possible, a valid conclusion to the given premises.

18. All dogs are barkers.
All cats are meowers.
No meowers are barkers.

19. All math classes are fun.
All fun things are interesting.

20. All sunny days are enjoyable.
Rainy days are not enjoyable.

CHAPTER 12

ANALYTIC GEOMETRY

SUPPLEMENTARY EXERCISES

1. Find the equation of the line with slope of 7 and passing through $(2, -1)$.
$(y = 7x - 15)$

2. Find the equation of the line with slope of -5 and passing through the origin.
$(y = -5x)$

3. Find the equation of the line passing through the origin and $(-3, -5)$.
$\left(y = \frac{5}{3}x\right)$

4. Find the equation of the line passing through $(0, 3)$ and $(3, 0)$.
$(y = -x + 3)$

5. Find the equation of the line parallel to the graph of $3x + 2y = 12$ and passing through the origin. $\left(y = -\frac{3}{2}x\right)$

6. Find the equation of the line perpendicular to the graph of $3x + 2y = 12$ and passing through $(-5, 4)$. $\left(y = \frac{2}{3}x + \frac{22}{3}\right)$

7. Find the equation of the line parallel to the graph of $3x + 6y = 14$ with the same x-intercept as the line $3x + 4y = 12$. $(y = -\frac{1}{2}x + 2)$

8. Find the equation of the line parallel to the y-axis passing through the point $(2, 3)$.
$(x = 2)$

9. Find the undirected distance between $(2, 5)$ and $(8, 13)$. (10)

10. Find the undirected distance between $(-3, -5)$ and $(9, -10)$. (13)

Using analytic geometry techniques, prove the following.

11. The median of a trapezoid is parallel to the bases and its length is half the sum of the lengths of the bases.

12. A triangle with two equal medians is isosceles.

13. The medians of a triangle intersect at a point two-thirds of the way from each vertex to the opposite side.

14. The sum of the squares of the lengths of the diagonals of a parallelogram is equal to the sum of the squares of the lengths of the sides.

15. If the diagonals of a quadrilateral bisect each other, the quadrilateral is a parallelogram.

16. The diagonals of a rhombus are perpendicular to each other.

Chapter 12 Test

Form 1

Decide if each of the following statements is true or is false. Then mark T or F in the space provided.

________ 1. The point of intersection of the coordinate axes is called the origin.

________ 2. The first number of an ordered pair of coordinates is the y-coordinate.

________ 3. The coordinate axes divide the plane in four quadrants.

________ 4. The length of $\overline{AB}$ parallel to the x-axis is found by subtracting the x-coordinate of B from the x-coordinate of A.

________ 5. An equation of the form $Ax + By + C = 0$ is called a linear equation.

________ 6. If P(2, 4) and Q(3, −5), the midpoint of $\overline{PQ}$ has coordinates of $\left(\frac{5}{2}, \frac{9}{2}\right)$.

________ 7. If two distinct lines have the same slope, they are parallel.

________ 8. If the slopes of two distinct lines are reciprocals, the lines are perpendicular.

________ 9. Lines parallel to the y-axis have no defined slope.

________ 10. The graph of $3x - 4y = 0$ passes through the origin.

Solve each problem and place the answer in the space provided.

________ 11. Find the length of $\overline{PQ}$ if P(8, −8) and Q(−1, 4).

________ 12. Find the midpoint of $\overline{MN}$ if M(8, 7) and N(−2, 8).

________ 13. Find the slope of $\overline{PQ}$ if P(8, 10) and Q(2, −5).

__________14. Find the equation of the line passing through (1, 1) with a slope of $\frac{1}{2}$.

__________15. Find the equation of the line passing through $(-4, 5)$ and (8, 12).

__________16. Find the slope of the line $3x - 2y = 12$. (Hint: Find two points on the line.)

__________17. Write the equation of the line parallel to the x-axis and 4 units above it.

__________18. Give the slope of a line perpendicular to a line with slope of $\frac{3}{4}$.

__________19. Write the equation of a line perpendicular to the x-axis containing the point (8, 10).

__________20. Solve the system

$$\begin{cases} x - 3y = 1 \\ x + 5y = 9 \end{cases}$$

21. Use the distance formula to show that $\triangle ABC$ is isosceles if A(2, 0), B(8, 10), and C(0, 8).

Chapter 12 Test

Form 2

Decide if each of the following statements is true or is false. Then mark T or F in the space provided.

________ 1. The x-coordinate of $(-3, 2)$ is -3.

________ 2. The point $(4, -5)$ lies in Quadrant II.

________ 3. If (x, y) is a point in Quadrant IV, then x is negative and y is positive.

________ 4. A point on the x-axis has a y-coordinate of 0.

________ 5. The length of a line segment $\overline{AB}$ parallel to the y-axis is found by subtracting the x-coordinate of B from the x-coordinate of A.

________ 6. If $P(-1, 3)$ and $Q(3, 5)$, the midpoint of $\overline{PQ}$ has coordinates of $(1, 4)$.

________ 7. The points $(5, 1)$ and $(5, -7)$ lie on a vertical line.

________ 8. The length of the line segment joining the points $(5, -1)$ and $(10, 11)$ is 13.

________ 9. The slope of the line passing through the points $P(5, -3)$ and $Q(-3, 7)$ is $-\frac{5}{4}$.

________ 10. The lines $x + 2y = 6$ and $2x + y = 5$ are perpendicular.

Solve each problem and place the answer in the space provided.

________ 11. Find the length of $\overline{AB}$ if $A(-4, 3)$ and $B(-1, 7)$.

________ 12. Find the midpoint of $\overline{AB}$ if $A(-4, 2)$ and $B(-2, 8)$.

________ 13. Find the slope of $\overline{PQ}$ if $P(5, -6)$ and $Q(-3, 2)$.

________ 14. Find the slope of the line $3x + 2y = -12$. (Hint: Find two points on the line.)

_________15. Find the equation of the line passing through (4, 6) with a slope of $\frac{3}{2}$.

_________16. Find the equation of the line passing through (5, 2) and (9, 6).

_________17. Find the equation of the line that passes through the point (3, 4) and that is parallel to $y = 2x + 1$.

_________18. Give the slope of a line perpendicular to the line $y = \frac{2}{3}x + 4$.

_________19. Write the equation of a line perpendicular to the y-axis containing the point $(-6, 3)$.

_________20. Solve the system

$$\begin{cases} 4x + 2y = -2 \\ x + 2y = -11 \end{cases}$$

21. Use the distance formula to show that trapezoid ABCD is isosceles if A(0, 0), B(10, 0), C(2, 8), and D(8, 8).

Chapter 12 Test

Form 3

Decide if each of the following statements is true or is false. Then mark T or F in the space provided.

________ 1. The points $(-5, 3)$ and $(3, -5)$ lie in the same quadrant.

________ 2. The point $(-6, 5)$ lies in Quadrant II.

________ 3. If (x, y) is a point in Quadrant III, then x is negative and y is negative.

________ 4. Points $(6, 1)$ and $(6, 4)$ lie on a horizontal line.

________ 5. The length of a line segment $\overline{AB}$ joining points $A(-5, 0)$ and $B(0, 12)$ is 17.

________ 6. If $P(5, 6)$ and $Q(-1, 4)$, the midpoint of $\overline{PQ}$ has coordinates of $(3, 1)$.

________ 7. The graph of $x = 7$ is a vertical line.

________ 8. The lines $2y = 4x + 10$ and $-2x + y = 5$ are parallel.

________ 9. If the product of the slopes of two distinct lines is 1, the lines are parallel.

________ 10. The y-intercept of $2x + 3y = 21$ is 7.

Solve each problem and place the answer in the space provided.

________ 11. Find the length of $\overline{AB}$ if $A(4, 19)$ and $B(-1, 7)$.

________ 12. Find the midpoint of $\overline{AB}$ if $A(-5, 6)$ and $B(-1, 8)$.

________ 13. Find the slope of $\overline{PQ}$ if $P(9, -4)$ and $Q(-1, 3)$.

________ 14. Find the slope of the line $4x + 7y = 28$. (Hint: Find two points on the line.)

________15. Find the equation of the line passing through (4, 5) with a slope of $\frac{3}{4}$.

________16. Find the equation of the line passing through (4, 1) and (9, 6).

________17. Find the equation of the line that passes through the point (2, 7) and that is parallel to $y = 4x + 8$.

________18. Give the slope of a line perpendicular to the line $y = -\frac{6}{7}x + 4$.

________19. Write the equation of a line perpendicular to the x-axis containing the point (1, 9).

________20. Solve the system

$$\begin{cases} 2x - 3y = 3 \\ -2x + 5y = -1 \end{cases}$$

21. Use the distance formula to show that $\triangle ABC$ is equilateral if $A(-4, 0)$, $B(4, 0)$, and $C(0, 4\sqrt{3})$.

Final Examination

Form 1

(2 Hours)

Decide if each of the following statements is true or is false. Then mark T or F in the space provided.

________ 1. Complementary angles are two angles in which the sum of their measures is 180°.

________ 2. Perpendicular lines intersect and form right angles.

________ 3. An acute angle has a measure of less than 90°.

________ 4. Adjacent angles have a common vertex but no common sides.

________ 5. For any real numbers a, b, c, if $a > b$, and $b > c$, then $a > c$.

________ 6. For any real numbers (including 0) a, b, c, if $ab = ac$, then $b = c$.

________ 7. Through two points one and only one straight line exists.

________ 8. The sum of the measures of two acute angles can be equal to the sum of an obtuse angle.

________ 9. A hexagon is a polygon with six sides.

________10. A straight angle has measure of 180°.

________11. $p \Rightarrow q$ symbolically represents a conditional statement.

________12. The converse of $p \Rightarrow q$ is $q \Rightarrow p$.

________13. All regular polygons have an apothem.

________14. A scalene triangle is a triangle with all sides unequal and all angles unequal.

________15. Two triangles are congruent if SSA $\cong$ SSA.

________16. The median of a triangle bisects one side and bisects one angle of a triangle.

________17. Vertical angles are always equal.

________18. Parallel lines intersect and bisect each other.

________19. In a given plane, through any point on a straight line there exists one and only one perpendicular to the line.

________20. If two lines are perpendicular to the same line, they are perpendicular to each other.

________21. A transversal is a line which intersects two or more straight lines.

________22. All three vertices of a triangle cannot be on the same line.

________23. Only one angle of a triangle can be a right angle or an obtuse angle.

_________24. The acute angles of a right triangle are supplementary.

_________25. If the measure of one acute angle of a right triangle is 60°, the side opposite this angle is one-half the length of the hypotenuse.

_________26. If two angles of a triangle are equal, the triangle is isosceles.

_________27. A square is not a rhombus.

_________28. A pentagon has nine sides.

_________29. Any two consecutive angles of a parallelogram are complementary.

_________30. Either diagonal of a parallelogram divides it into two congruent triangles.

_________31. If both sets of opposite sides of a quadrilateral are equal, then the quadrilateral is a parallelogram.

_________32. The segment joining the midpoints of two sides of a triangle is parallel to the base.

_________33. Only in an isosceles triangle is the measure of the median equal to one-half the measure of the base.

_________34. A chord is a line segment which begins outside of a circle and passes through it.

_________35. Tangent circles have equal radii.

_________36. The measure of an arc is defined as the measure of the central angle which intercepts it.

_________37. An angle inscribed in a semicircle is a right angle.

_________38. All angles intercepting the same arc are equal.

_________39. In the same or congruent circles, equal chords have equal arcs.

_________40. A line through the center of a circle and perpendicular to a chord bisects the chord.

_________41. In a circle, equal chords are equidistant from the center.

_________42. A tangent intersects a circle at only one point.

_________43. A secant of a circle contains a chord of the circle.

_________44. If a, b, and c, are real numbers, and $a < b, b < c$, then $c < a$.

_________45. No two angles of an equilateral triangle have the same measure.

_________46. The diagonals of a rhombus that is not a square are unequal.

_________47. Two rectangles which have equal perimeters also have equal areas.

_________48. Doubling the radius of a circle will double its circumference.

_________49. The area of a circle equals $2\pi r^2$.

_________50. A parallelogram is a rhombus.

_________51. If a parallelogram is inscribed in a circle, it must be a rectangle.

_________52. A trapezoid which has two equal sides is called an equilateral trapezoid.

_________53. The sum of the measures of the angles of a triangle equals 180°.

_________54. A postulate is a statement to be proved.

_________55. Angles of $(3x + 65)^\circ$ and $(115 - 3x)^\circ$ are supplementary.

_________56. An exterior angle of a triangle is equal to the sum of its nonadjacent interior angles.

_________57. The sum of the measures of two sides of a triangle is greater than the measure of the third side.

_________58. In a circle, the greater of the noncongruent arcs has the smaller chord.

_________59. Ratios are indicated quotients.

_________60. A proportion is the equality of two ratios.

_________61. In a proportion, the third and the fourth terms may be interchanged and the equality is undisturbed.

_________62. The slope of the line $2x + 3y = 12$ is $-\frac{2}{3}$.

_________63. All congruent figures are also similar figures.

_________64. If two triangles have two pair of corresponding sides in proportion, they are similar.

_________65. The square of the length of the hypotenuse of a right triangle is equal to the difference of the squares of the lengths of the legs.

_________66. A locus is the set of all points which satisfy a given condition and only those points.

_________67. The locus of all points a given distance from a fixed point is a circle whose diameter is the given distance.

_________68. In a right triangle, $\sin A = \frac{\text{measure of the side opposite } \angle A}{\text{measure of the side adjacent } \angle A}$.

_________69. If $\cos A = \frac{4}{5}$, then $\sin A = \frac{3}{5}$.

_________70. It has be proven impossible to bisect an angle using only a compass, a straightedge, and a pencil.

_________71. It is impossible to circumscribe a circle about a triangle using a compass, a straightedge, and a pencil.

_________72. The area of a trapezoid is equal to one-half the product of the altitude and the sum of its bases.

________73. In a right triangle, $\cos A = \frac{\text{measure of the side adjacent } \angle A}{\text{measure of the hypotenuse}}$.

________74. All equilateral triangles are regular polygons.

________75. The complement of an angle is always an acute angle.

Solve each problem and place the answer in the space provided.

________76. $m(\widehat{ADC}) =$

________77. $m(\angle ABC) =$

________78. $m(\overline{AOC}) =$

________79. $m(\overline{OB}) =$

________80. Find the measure of the circumference.

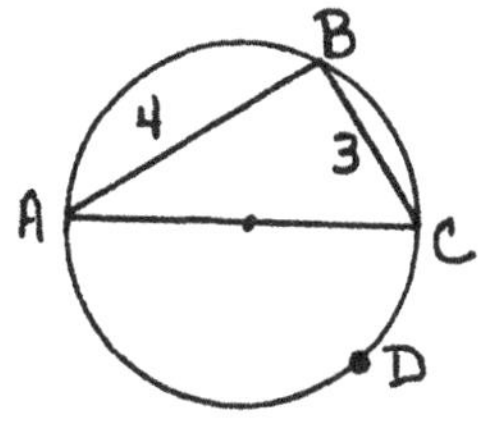

(Questions 76–80)

________81. $m(\overline{BC}) =$

________82. $m(\overline{AD}) =$

________83. $m(\overline{AC}) =$

________84. $m(\angle DBC) =$

________85. Area of $\triangle ABC =$

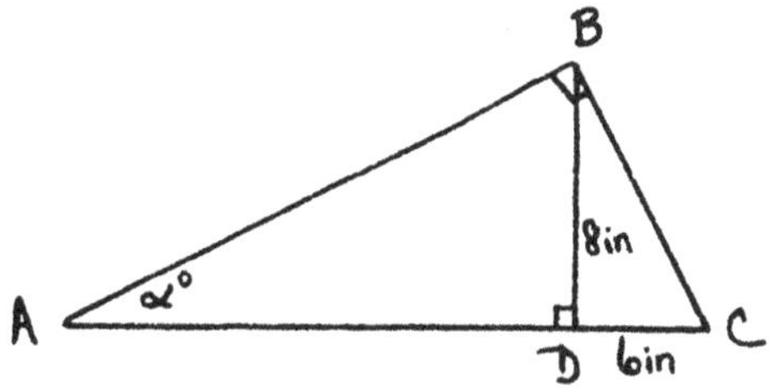

(Questions 81–85)

________86. $m(\widehat{AC}) =$

________87. $m(\angle ABC) =$

________88. $m(\angle GAB) =$

________89. $m(\widehat{DC}) =$

________90. $m(\angle DGC) =$

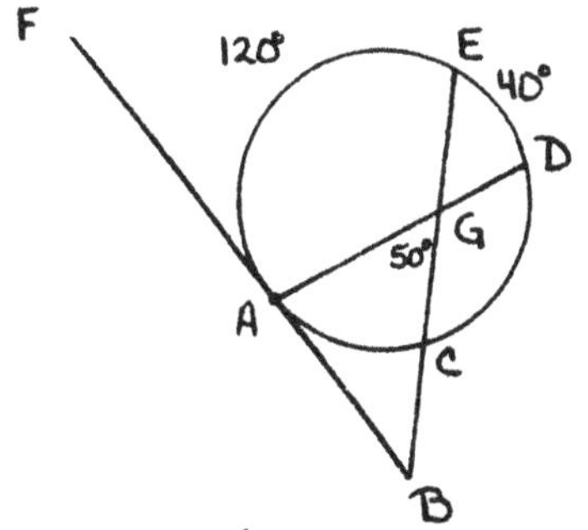

(Questions 86–90)

ABDEF is a regular pentagon with center at G.

_________91. m(∠DGC) =

_________92. m(∠FED) =

_________93. m(∠GDC) =

_________94. m(∠FEH) =

_________95. m($\overline{GD}$) =

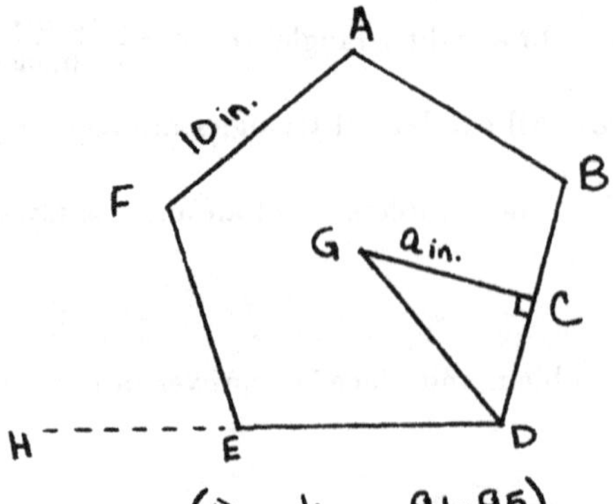

(Questions 91-95)

_________96. m($\widehat{AC}$) =

_________97. m($\widehat{BC}$) =

_________98. m(∠BAC) =

_________99. m(∠BOC) =

_________100. The area of the circle is?

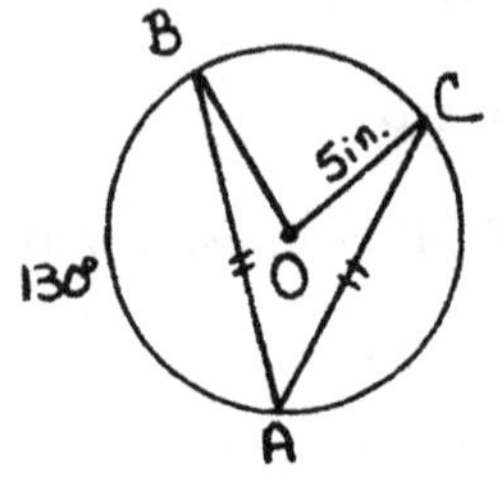

(Questions 96-100)

PROOFS

1) Prove that the sum of the measures of the angles of a triangle is 180°.

2) The length of the perimeter of a quadrilateral is greater than the sum of the lengths of its diagonals.

Final Examination

Form 2

(2 Hours)

Decide if each of the following statements is true or is false. Then mark T or F in the space provided.

_________ 1. Supplementary angles are two angles in which the sum of their measures is 180°.

_________ 2. The complement of an angle is always an acute angle.

_________ 3. An obtuse angle has a measure of less than 90°.

_________ 4. Adjactent angles can be vertical angles.

_________ 5. Angles of $(4x + 125)^\circ$ and $(55 - 4x)^\circ$ are complementary.

_________ 6. A bisector of a right angle divides the angle into two complementary angles.

_________ 7. The converse of a true conditional statement is always true.

_________ 8. If three angles of one triangle are congruent to three angles of another triangle, the triangles are congruent.

_________ 9. If two angles in a triangle are complementary, the triangle is a right triangle.

_________10. The bisector of the vertex angle of an isosceles triangle is perpendicular to the base.

_________11. An equilateral triangle is also isosceles.

_________12. If in $\triangle ABC$ and $\triangle DEF$, $\angle A \cong \angle D$, $\angle B \cong \angle E$, and $\overline{AB} \cong \overline{DE}$, then $\triangle ABC \cong \triangle DEF$.

_________13. Alternate interior angles are always congruent.

_________14. The sum of the measures of the exterior angles of a hexagon is 720°.

_________15. The sum of the measures of the exterior angles of a triangle is equal to twice the sum of the measures of the interior angles of the triangle.

_________16. The sum of the measures of the interior angles of a pentagon is 540°.

_________17. A triangle can have both a right angle and an obtuse angle.

_________18. The measure of an interior angle of a regular hexagon is 120°.

_________19. If two isosceles triangles are congruent, then their vertex angles are congruent.

_________20. If two lines are cut by a transversal such that alternate exterior angles are congruent, the lines are parallel.

_________21. The measure of an exterior angle of a triangle is equal to the sum of the measures of the nonadjacent interior angles.

_________22. The acute angles of a right triangle are supplementary.

_________23. The diagonal of a parallelogram divides the parallelogram into two congruent triangles.

_________24. Consecutive angles of a quadrilateral are supplementary.

_________25. A rhombus is a square.

_________26. The diagonals of a rhombus bisect each other.

_________27. The diagonals of a rectangle are perpendicualr.

_________28. Consecutive angles of a rhombus are supplementary.

_________29. If the base angles of a trapezoid are congruent, the trapezoid is isosceles.

_________30. The distance between two parallel lines depends on the slant of the transversal.

_________31. The area of a parallelogram with a base measuring 4 cm and a height of 7 cm is 28 sq. cm.

_________32. The area of a circle with a diameter of 8 in is 16π sq. in.

_________33. In a $30^\circ - 60^\circ$ right triangle, the length of the side opposite the 30° angle is one-half the length of the hypotenuse.

_________34. The volume of a rectangular solid measuring 3 ft by 2 ft by 5 ft is 15 sq. ft.

_________35. The radius of a circle with area of 49π sq m is 7 m.

_________36. The area of a rhombus with diagonals that are 5 cm and 8 cm long is 40 sq cm.

_________37. A line parallel to one side of a triangle divides the other two sides proportionally.

_________38. All pentagons are similar.

_________39. If $\frac{a}{b} = \frac{c}{d}$, then $\frac{a-b}{b} = \frac{c-d}{d}$.

_________40. The measures of all corresponding sides of similar triangles are in proportion.

_________41. The fourth proportional of 5, 11, and 15 is 33.

_________42. The center of a circle is a point of the circle.

_________43. A perpendicular bisector of a chord of a circle always passes through the center of the circle.

_________44. An inscribed angle can never be a right angle.

_________45. In a circle, perpendicular lines intersect congruent arcs.

_________46. Secants drawn to a circle from a point outside the circle are congruent.

_________47. If two diameters intersect within a circle, their point of intersection is the center of the circle.

_________48. The relation "$<$" is a transitive relation.

_________49. The largest side of a triangle is always opposite the largest angle.

_________50. In a triangle, an exterior angle is greater than each interior angle.

_________51. If $a < b$ and $c < d$, then $a + c < b + d$.

_________52. The measure of the hypotenuse of a right triangle is greater than the sum of the measures of the legs.

_________53. If unequal quantities are multiplied to unequal quantities, then the products are unequal quantities.

_________54. A chord can be greater than a diameter of the circle.

_________55. If a central angle has a measure of 67°, then the chord intercepted by the angle is longer than the radius of the circle.

_________56. The longer of two chords in a circle is nearer to the center of the circle.

_________57. All bisectors of the three sides of a triangle meet in a point.

_________58. A square can be inscribed within a circle with a compass and a striaghtedge.

_________59. The medians of a triangle always meet at a point inside the triangle.

_________60. A circle can be constructed through any three noncollinear points.

_________61. The locus of all points equidistant from two points is any bisector of the segment joining the two points.

_________62. The circumference of a circle is given by the formula $C = d\pi$.

_________63. If a circle has a radius of 2 in, its area is numerically equal to its circumference.

_________64. A circle can be inscribed in a rectangle that is not a square.

_________65. If a regular hexagon (six-sided figure) has sides of 6 cm and its apothem is $3\sqrt{3}$ cm, the area of the hexagon is $18\sqrt{3}$ sq cm.

_________66. The circumference of a circle with diameter of 7 in long is 14π in.

_________67. If p is false and q is true, the $\sim p \wedge q$ is true.

_________68. In a right triangle, $\sin A = \dfrac{\text{measure of the side opposite } \angle A}{\text{measure of the hypotenuse}}$.

_________69. The contrapositive of $\sim p \Rightarrow q$ is $\sim q \Rightarrow p$.

_________70. A statement and it converse are logically equivalent.

_________71. The length of $\overline{AB}$ parallel to the x-axis is found by subtractin the x-coordinate of A from the x-coordinate of B.

________72. If P(3, 1) and Q(9, 11), the midpoint of $\overline{PQ}$ has coodinated of (3, 5).

________73. If two distinct lines have the same slope, the lines are perpendicular.

________74. The points (5, 2) and (5, 6) lie on a vertical line.

________75. If the product of the slopes of two distinct lines is -1, the lines are perpendicular.

Solve each problem and place the answer in the space provided.

________76. $m(\widehat{ADC}) =$

________77. $m(\angle ABC) =$

________78. $m(\overline{AOC}) =$

________79. $m(\overline{OB}) =$

________80. Find the area of the circle.

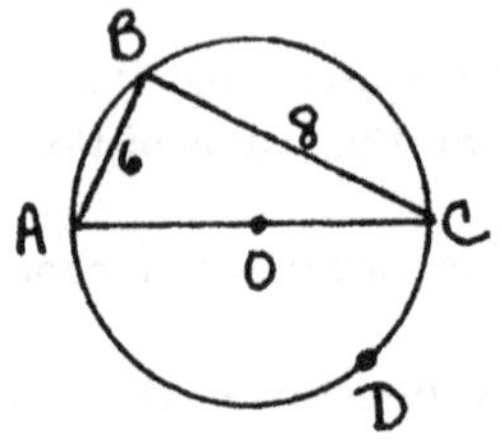

(Questions 76-80)

________81. $m(\overline{BC}) =$

________82. $m(\overline{AD}) =$

________83. $m(\overline{AC}) =$

________84. $m(\angle DBC) =$

________85. Area of $\triangle ABC =$

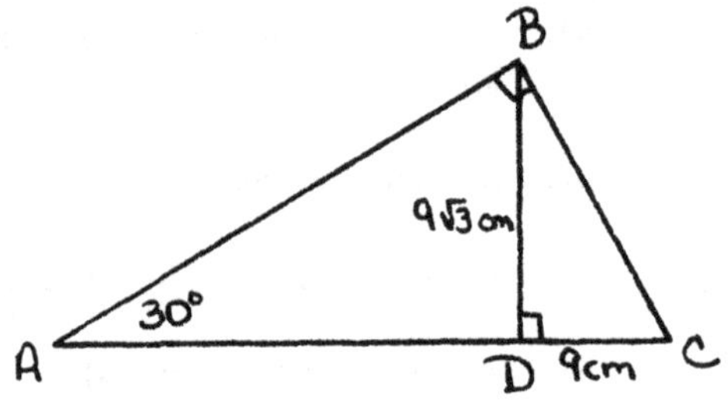

(Questions 81-85)

________86. $m(\widehat{AC}) =$

________87. $m(\angle ABC) =$

________88. $m(\angle GAB) =$

________89. $m(\widehat{DC}) =$

________90. $m(\angle DGC) =$

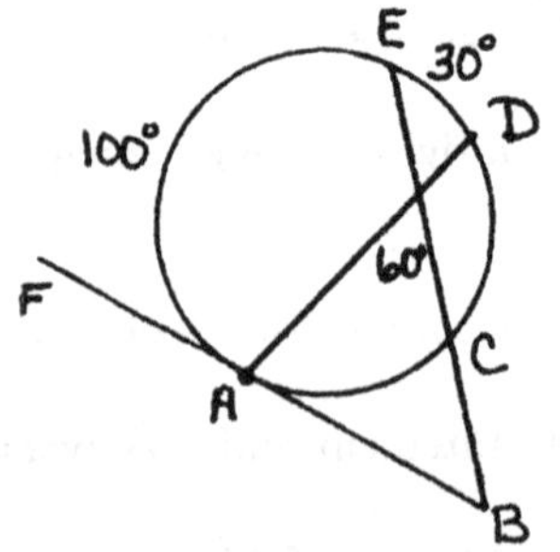

(Questions 86-90)

ABDEF is a regular pentagon with center at G.

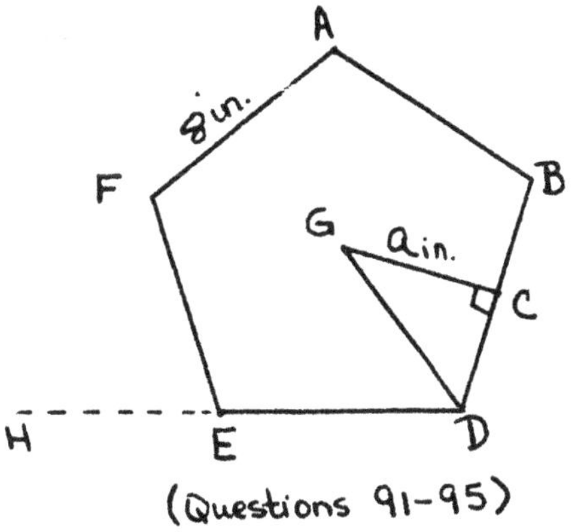

________91. m(∠DGC) =

________92. m(∠FED) =

________93. m(∠GDC) =

________94. m(∠FEH) =

________95. Find the area of the pentagon.

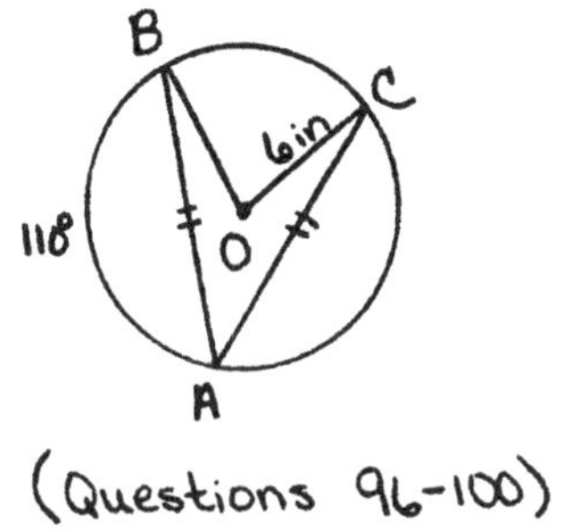

________96. $m(\widehat{AC}) =$

________97. $m(\widehat{BC}) =$

________98. m(∠BAC) =

________99. m(∠BOC) =

________100. The area of the circle is?

PROOFS

1) Prove that the sum of the measures of the angles of a hexagon is 720°.

2) Given: △AFB ≅ △DCE.

$\overline{FD} \parallel \overline{AC}$.

Prove: $\overline{FB} \parallel \overline{EC}$.

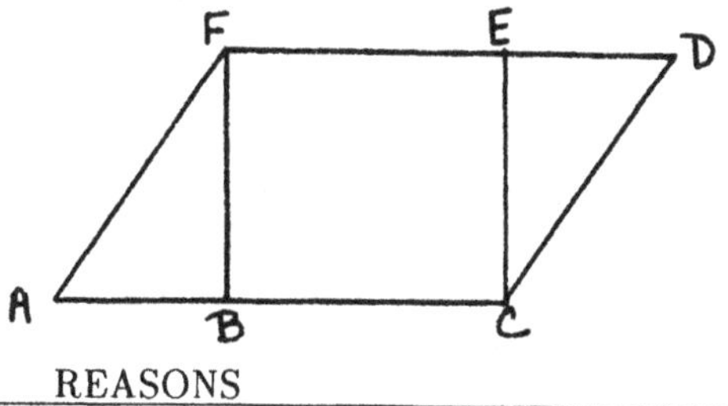

STATEMENTS	REASONS

Final Examination

Form 3

(2 Hours)

Decide if each of the following statements is true or is false. Then mark T or F in the space provided.

_________ 1. Supplementary angles are two angles in which the sum of their measures is 90°.

_________ 2. A scalene triangle has a pair of congruent sides.

_________ 3. An angle of 135° is an obtuse angle.

_________ 4. There can be only one perpendicular to a line through a given point on the line.

_________ 5. A theorem is a statement that has been proved.

_________ 6. If angles are congruent, they are vertical angles.

_________ 7. A line segment has only one perpendicular bisector.

_________ 8. The sum of the measures of two acute angles can be equal to the sum of an obtuse angle.

_________ 9. An angle of 27° has no supplement.

_________ 10. An obtuse triangle has only 2 altitudes.

_________ 11. Angles of $(3x + 154)^\circ$ and $(36 - 3x)^\circ$ are supplementary.

_________ 12. If angles are vertical, they are congruent.

_________ 13. An angle of 93° has no complement.

_________ 14. A scalene triangle is a triangle with all sides unequal and all angles unequal.

_________ 15. Two triangles are congruent if $AAA \cong AAA$.

_________ 16. The bisector of the vertex angle of an isosceles triangle bisects the base of the triangle.

_________ 17. Two right triangles are congruent is they have equal hypotenuses.

_________ 18. All equilateral triangles are congruent.

_________ 19. If two parallel lines are cut by a transversal, then interior angles on the same side of the transversal are supplementary.

_________ 20. If two lines are perpendicular to the same line, they are parallel to each other.

_________ 21. If two angles in a triangle are 36° and 24°, then the triangle is an obtuse triangle.

_________ 22. An interior angle of a regular octagon (eight-sides) has measure 135°.

_________ 23. The measure of an exterior angle of a triangle is supplementary to the sum of the measures of the nonadjacent interior angles.

_________24. If two parallel lines are cut by a transversal, then interior angles on the same side of the transversal can never be complementary.

_________25. Through a point not on the line there exists infinitely many parallel lines to the line.

_________26. If a triangle has an obtuse angle, the other two angles must be acute angles.

_________27. The sum of the measures of the exterior angles of an octagon is 360°.

_________28. If two lines are cut by a transversal, alternate interior angles are congruent.

_________29. Opposite angles of a parallelogram are congruent.

_________30. Either diagonal of a rhombus divides it into two congruent triangles.

_________31. If both sets of opposite sides of a quadrilateral are equal, then the quadrilateral is a parallelogram.

_________32. The segment joining the midpoints of two sides of a triangle is parallel to the base.

_________33. A parallelogram is a trapezoid.

_________34. In a $30^\circ - 60^\circ$ right triangle, the side opposite the 60° angle is twice as long as the side opposite the 30° angle.

_________35. The volume of a rectangular solid measuring 3 in by 5 in by 4 in is 30 cubic in.

_________36. A triangle with sides measuring 10 cm, 24 cm and 26 cm is a right triangle.

_________37. The area of a rhombus with diagonals that are 4 cm and 7 cm long is 28 sq cm.

_________38. The area of a circle with diameter that is 16 in. long is 64π sq in.

_________39. The fourth proportional of 6, 12, and 18 is 24.

_________40. If $\cos A = \frac{5}{13}$, then $\sin A = \frac{12}{13}$.

_________41. If two right triangles have a respective pair of acute angles congruent, the triangles are similar.

_________42. All regular hexagons are similar.

_________43. All parallelograms are similar.

_________44. In a proportion, the denominators of the two ratios are equal.

_________45. All isosceles right triangles are similar.

_________46. Sides of similar triangles have equal measure.

_________47. A line that intersects a circle in exactly one point is called a tangent to the circle.

_________48. Doubling the radius of a circle will double its circumference.

________49. A central angle of a circle has half as many degrees as its intercepted arc.

________50. A bisector of a chord of a circle always passes through the center of the circle.

________51. A radius is a chord of a circle.

________52. The measure of an inscribed angle is equal to one-half the measure of its intercepted arc.

________53. If two circles are tangent, then their point of tangency and the center of each circle are collinear.

________54. Opposite angles of an inscribed quadrilateral are congruent.

________55. The greater of two chords in a circle is nearer the center of the circle.

________56. In a triangle, an exterior angle is greater than each interior angle.

________57. If a right triangle is inscribed in a circle, the hypotenuse of the triangle is a diameter of the circle.

________58. In congruent circles, the greater of two minor arcs has the shorter chord.

________59. If a central angle of a circle has a measure of 90°, then the chord intercepted by the angle is longer than a radius of the circle.

________60. The medians of a triangle always meet at a point inside the triangle.

________61. An angle bisector is the locus of all points equidistant from the sides of an angle.

________62. A circumscribed circle is a circle inside a triangle which is tangent to the three sides of the triangle.

________63. It is always possible to construct a circle passing through four points in a plane.

________64. If a square has sides of 6 cm and an apothem of a cm, its area is $12a$ sq cm.

________65. An apothem of a regular polygon is a line segment drawn from the center of the polygon perpendicular to one of the polygon's sides.

________66. The central angle of a regular octagon has measure of 60°.

________67. The apothem of a regular polygon is the radius of the inscribed circle of the polygon.

________68. In a right triangle, $\sin A = \dfrac{\text{measure of the side opposite } \angle A}{\text{measure of the hypotenuse}}$.

________69. The contrapositive of the contrapositive of a statement is the original statement.

________70. The inverse of $p \Rightarrow \sim q$ is $\sim q \Rightarrow p$.

________71. If p is false and q is true then $\sim (p \vee q)$ is false.

_________72. The point $(-3, 4)$ is in the fourth quadrant.

_________73. The midpoint of the line segment connecting the points P(2, 6) and Q$(-6, 8)$ is $(-2, 7)$.

_________74. The slope of $3x + 4y = 12$ is 3.

_________75. Lines parallel to the y-axis have no defined slope.

Solve each problem and place the answer in the space provided.

_________76. $m(\overset{\frown}{CD}) =$

_________77. $m(\angle AEB) =$

_________78. $m(\angle B) =$

_________79. $m(\angle C) =$

_________80. $m(\overset{\frown}{BC}) =$

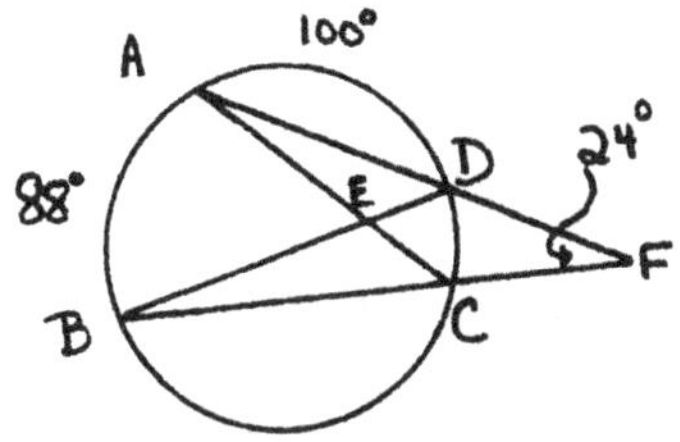

(Questions 76-80)

In the figure, D and E are midpoints of their respective sides.

_________81. $m(\overline{BC}) =$

_________82. $m(\overline{AB}) =$

_________83. Find the perimeter of $\triangle ABC$

_________84. $m(\angle A) =$

_________85. Area of $\triangle ABC =$

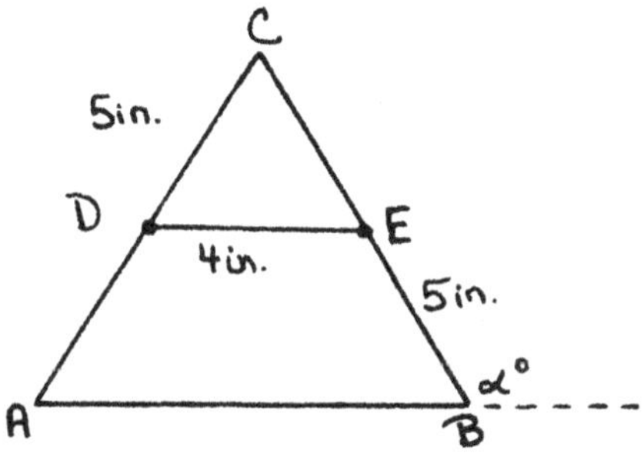

(Questions 81-85)

_________86. $m(\overset{\frown}{BC}) =$

_________87. $m(\angle A) =$

_________88. $m(\overline{AOB}) =$

_________89. $m(\overline{OC}) =$

_________90. Find the area of the circle O.

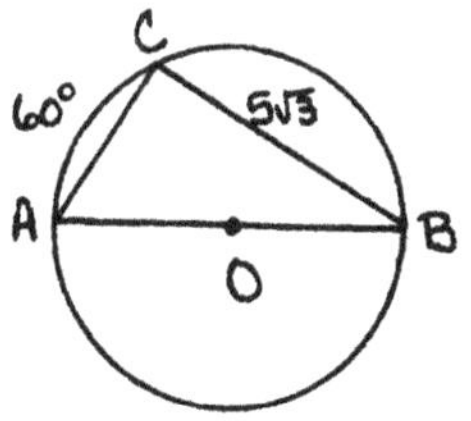

(Questions 86-90)

ABDEF is a regular hexagon with center at G.

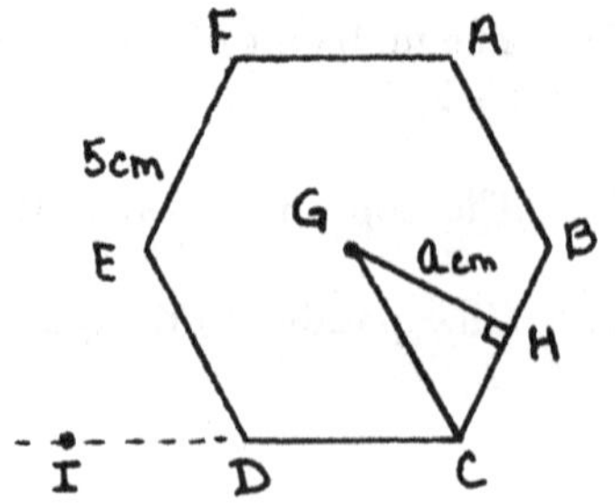

________91. m(∠CGH) =

________92. m(∠EDC) =

________93. m(∠GCH) =

________94. m(∠EDI) =

________95. Find the area of the hexagon.

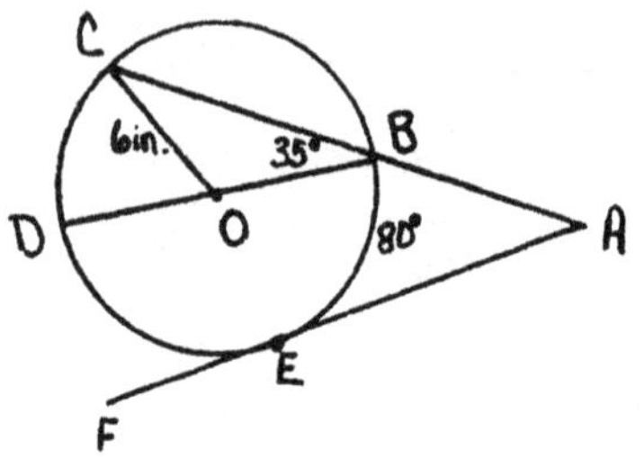

________96. m($\overset{\frown}{CD}$) =

________97. m($\overset{\frown}{BC}$) =

________98. m(∠BCO) =

________99. m(∠A) =

________100. Find the circumference of the circle.

PROOFS

1) Prove that the sum of the measures of the angles of an octagon is 1080°.

2) Prove: If the base angles of a trapezoid are congruent, then the trapezoid is an isosceles trapezoid.

ANSWERS TO CHAPTER TESTS

CHAPTER 1 TEST
FORM 1

1.	T
2.	F
3.	T
4.	F
5.	T
6.	T
7.	F
8.	T
9.	T
10.	T
11.	F
12.	F
13.	T
14.	F
15.	T
16.	3
17.	2
18.	123°
19.	3°
20.	13 cm
21.	18 in.
22.	6 m
23.	57°
24.	8 in.
25.	62°

CHAPTER 1 TEST
FORM 2

1.	F
2.	F
3.	T
4.	T
5.	T
6.	F
7.	F
8.	F
9.	T
10.	T
11.	T
12.	F
13.	T
14.	F
15.	T
16.	7
17.	20
18.	33°
19.	137°
20.	28 cm
21.	6 m
22.	7 cm
23.	20
24.	7 in.
25.	52°

CHAPTER 1 TEST
FORM 3

1.	T
2.	F
3.	F
4.	T
5.	T
6.	T
7.	F
8.	T
9.	T
10.	F
11.	F
12.	T
13.	T
14.	T
15.	F
16.	11
17.	9
18.	66°
19.	35°
20.	15 cm
21.	5 m
22.	2 cm
23.	10
24.	160°
25.	168°

CHAPTER 2 TEST
FORM 1

1.	T
2.	T
3.	F
4.	T
5.	F
6.	T
7.	T
8.	T
9.	F
10.	T
11.	T
12.	F
13.	T
14.	F
15.	F
16.	50°
17.	4 in.
18.	40°
19.	$5\frac{1}{2}$ m
20.	21 cm

CHAPTER 2 TEST
FORM 2

1.	T
2.	F
3.	T
4.	F
5.	F
6.	T
7.	F
8.	T
9.	T
10.	F
11.	T
12.	T
13.	T
14.	F
15.	T
16.	100°
17.	10 cm
18.	130°
19.	60
20.	30 m

CHAPTER 2 TEST
FORM 3

1.	F
2.	T
3.	T
4.	F
5.	F
6.	T
7.	F
8.	F
9.	T
10.	F
11.	T
12.	T
13.	F
14.	F
15.	T
16.	60°
17.	60°
18.	70°
19.	80
20.	27 cm

CHAPTER 3 TEST
FORM 1

1. T
2. F
3. F
4. T
5. F
6. T
7. T
8. F
9. T
10. F
11. T
12. F
13. T
14. F
15. T
16. 30°
17. 150°
18. 110°
19. 900°
20. 108°
21. 34°
22. No
23. 10°
24. 70°

CHAPTER 3 TEST
FORM 2

1. F
2. F
3. T
4. T
5. T
6. F
7. T
8. T
9. F
10. F
11. F
12. T
13. F
14. T
15. F
16. 60°
17. 120°
18. 125°
19. 1260°
20. $128\frac{4}{7}°$
21. Yes
22. 50°
23. 50°
24. 140°

CHAPTER 3 TEST
FORM 3

1. F
2. T
3. T
4. F
5. F
6. F
7. T
8. F
9. T
10. T
11. F
12. T
13. T
14. T
15. T
16. 130°
17. 50°
18. 115°
19. 1080°
20. 120°
21. No
22. 75°
23. 70°
24. 70°

CHAPTER 4 TEST
FORM 1

1. T
2. F
3. F
4. F
5. F
6. T
7. F
8. T
9. T
10. F
11. T
12. T
13. F
14. T
15. T
16. 75°
17. 12 cm
18. 15 m
19. 60°
20. 46°
21. 56
22. $x = 1$
23. 150°

CHAPTER 4 TEST
FORM 2

1. F
2. F
3. F
4. T
5. T
6. F
7. T
8. T
9. T
10. T
11. F
12. F
13. T
14. T
15. T
16. 70°
17. 20 m
18. 22 cm
19. 140°
20. 35°
21. 129°
22. 80
23. 150°

CHAPTER 4 TEST
FORM 3

1. T
2. F
3. T
4. F
5. F
6. F
7. F
8. T
9. T
10. T
11. F
12. T
13. T
14. T
15. F
16. 57°
17. 6 cm
18. 16 m
19. 51°
20. 8 cm
21. 120°
22. 112
23. 60°

CHAPTER 5 TEST
FORM 1

1. T
2. F
3. F
4. F
5. T
6. T
7. T
8. F
9. F
10. F
11. 35 cm
12. 6 in
13. 6 ft
14. 218 sq ft
15. 12.5 sq yd
16. $\frac{8}{\sqrt{\pi}}$ cm
17. $100\sqrt{3}$ sq in
18. 25 sq ft
19. 26 m
20. 8 cm
21. 210 sq ft
22. 100π cu cm

CHAPTER 5 TEST
FORM 2

1. F
2. F
3. T
4. F
5. T
6. T
7. T
8. F
9. T
10. T
11. 96 sq cm
12. 9 cm
13. 2 in
14. 11 sq m
15. 32 sq yds
16. $\frac{10}{\pi}$ cm
17. $64\sqrt{3}$ sq in
18. 90 sq cm
19. 24 cm
20. 4 m
21. 72π sq m
22. 105 sq cm

CHAPTER 5 TEST
FORM 3

1. T
2. T
3. F
4. T
5. T
6. T
7. T
8. F
9. T
10. T
11. 66 sq m
12. 8 in
13. 2 cm
14. $\left(6\sqrt{3}+\frac{9}{2}\right)\text{in}^2$
15. 27 sq cm
16. 7 in
17. $36\sqrt{3}$ sq ft
18. 72 sq yd
19. 16 in
20. 2 m
21. 42π sq cm
22. $\frac{160\pi}{3}$ sq cm

CHAPTER 6 TEST
FORM 1

1. T
2. F
3. F
4. T
5. F
6. T
7. F
8. T
9. F
10. T
11. T
12. T
13. 8
14. $x = 2$
15. $x = 5$
16. 8
17. $\frac{7}{2}$
18. 12
19. $\frac{10}{3}$
20. $\frac{3}{2}$

CHAPTER 6 TEST
FORM 2

1. T
2. F
3. T
4. T
5. T
6. F
7. T
8. T
9. F
10. F
11. T
12. T
13. 8
14. $x = 4$
15. $x = 6$
16. 6
17. 7
18. 5
19. $\frac{4}{3}$
20. $\frac{5}{3}$

CHAPTER 6 TEST
FORM 3

1. T
2. F
3. F
4. T
5. T
6. F
7. T
8. F
9. T
10. T
11. T
12. T
13. 10
14. $x = 25$
15. $x = 20$
16. 6
17. 12
18. $\frac{7}{2}$
19. 36
20. $\frac{5}{12}$

Continued on the next page.

21. 30
22. ≈ 49 ft
23. $\frac{5}{13}$

CHAPTER 7 TEST FORM 1

1. F
2. T
3. F
4. T
5. T
6. F
7. T
8. T
9. F
10. F
11. 50°
12. 60°
13. 10°
14. 70°
15. 2
16. 100°
17. 40°
18. 90°
19. 140°
20. 40°

CHAPTER 8 TEST FORM 1

1. F
2. F
3. T
4. T
5. F
6. T
7. T
8. T
9. T
10. T
11. $x < 9$
12. $x > 9$
13. $x < \frac{7}{3}$
14. $x > 4$
15. $\angle A$
16. 60°
17. $\overline{AB}$
18. $\widehat{AB}$
19. $\overline{AC}$
20. Yes

21. 49
22. ≈ 17.6 m
23. $\frac{12}{5}$

CHAPTER 7 TEST FORM 2

1. F
2. T
3. T
4. F
5. F
6. F
7. T
8. F
9. F
10. T
11. 45°
12. 90°
13. 30°
14. 65°
15. 2
16. 40°
17. 20°
18. 100°
19. 100°
20. 80°

CHAPTER 8 TEST FORM 2

1. T
2. F
3. F
4. T
5. F
6. T
7. F
8. F
9. T
10. T
11. $x < 10$
12. $x > 16$
13. $x > 13$
14. $x > 14$
15. $\angle C$
16. $\angle 1$
17. $\overline{CD}$
18. $\widehat{AB}$
19. $\overline{AC}$
20. $\overline{CB}$

21. ≈ 18 ft
22. .5
23. $\frac{5}{13}$

CHAPTER 7 TEST FORM 3

1. T
2. T
3. T
4. T
5. T
6. F
7. F
8. T
9. T
10. F
11. 100°
12. 80°
13. 100°
14. 60°
15. 20
16. 30°
17. 140°
18. 10°
19. 100°
20. 50°

CHAPTER 8 TEST FORM 3

1. F
2. T
3. F
4. T
5. T
6. T
7. F
8. T
9. F
10. F
11. $x > 4$
12. $x > 30$
13. $x < 22$
14. $x > 21$
15. $\angle B$
16. $\overline{EF}$
17. $\overline{EF}$
18. $\overline{AB}$
19. $\overline{AC}$
20. No

CHAPTER 9 TEST
FORM 1

1.	T
2.	F
3.	T
4.	F
5.	F
6.	T
7.	T
8.	T
9.	F
10.	T
11.	T
12.	F
13.	T
14.	T
15.	T
16.	3 cm
17.	4 cm
18.	6 in
19.	a point
20.	Yes

CHAPTER 9 TEST
FORM 2

1.	T
2.	T
3.	F
4.	T
5.	F
6.	F
7.	T
8.	T
9.	F
10.	T
11.	T
12.	F
13.	T
14.	F
15.	T
16.	6 m
17.	15 m
18.	8 in
19.	4 in
20.	The center of the circle passing through the three points.

CHAPTER 9 TEST
FORM 3

1.	F
2.	T
3.	T
4.	T
5.	F
6.	T
7.	T
8.	F
9.	F
10.	T
11.	F
12.	T
13.	T
14.	T
15.	T
16.	8
17.	18
18.	10 cm
19.	3 m
20.	A circle

CHAPTER 10 TEST
FORM 1

1.	T
2.	T
3.	T
4.	F
5.	F
6.	T
7.	F
8.	T
9.	F
10.	T
11.	64π sq units
12.	6
13.	$\frac{25}{2}a$ sq units
14.	45°
15.	5 cm
16.	480 sq in
17.	No
18.	No
19.	9 in
20.	36°

CHAPTER 10 TEST
FORM 2

1.	F
2.	T
3.	T
4.	F
5.	T
6.	T
7.	T
8.	F
9.	F
10.	F
11.	10π m
12.	10
13.	$21a$ sq units
14.	72°
15.	7 m
16.	$96\sqrt{3}$ in
17.	No
18.	$\frac{4\pi}{3}$ sq units
19.	7 m
20.	4 cm

CHAPTER 10 TEST
FORM 3

1.	T
2.	F
3.	T
4.	T
5.	T
6.	F
7.	T
8.	T
9.	F
10.	T
11.	25π sq cm
12.	25 in
13.	$24a$ sq m
14.	36°
15.	8 cm
16.	$150\sqrt{3}$ sq in
17.	No
18.	16π sq cm
19.	16 m
20.	5 ft

CHAPTER 11 TEST
FORM 1

1. T	2. T	3. F	4. T	5. T
6. T	7. T	8. F	9. T	10. T

11. If Laurie isn't friendly, then Marcia isn't happy.
12. Laurie is friendly and, Marcia is happy or John is grouchy.
13. If a triangle doesn't have two congruent sides, then it doesn't have two congruent angles.
14.

p	q	$\sim p$	$\sim p \vee q$	$\sim q$	$p \wedge \sim q$	$\sim(p \wedge \sim q)$	$\sim p \vee q \Leftrightarrow \sim(p \wedge \sim q)$
T	T	F	T	F	F	T	T
T	F	F	F	T	T	F	T
F	T	T	T	F	F	T	T
F	F	T	T	T	F	T	T

15. No conclusion.
16. Joe is playing hooky.
17. I did not work hard.
18. No logic students are gullible.
19. No valid conclusion.
20. No rectangles are triangles.

CHAPTER 11 TEST
FORM 2

1. F	2. T	3. F	4. F	5. T
6. F	7. T	8. T	9. T	10. F

11. Laurie isn't friendly and Marcia isn't happy.
12. Laurie is friendly and Marcia is happy or John is not grouchy.
13. If a quadrilateral is not a rectangle, then it is not a square.
14.

p	q	$\sim p$	$\sim q$	$p \wedge q$	$\sim(p \wedge q)$	$\sim p \vee \sim q$	$\sim(p \wedge q) \Leftrightarrow \sim p \vee \sim q$
T	T	F	F	T	F	F	T
T	F	F	T	F	T	T	T
F	T	T	F	F	T	T	T
F	F	T	T	F	T	T	T

15. It did not rain.
16. Joe is sick.
17. No conclusion.
18. No bikes are cars.
19. All instructors are helpful.
20. No conclusion.

CHAPTER 11 TEST
FORM 3

1. T	2. F	3. F	4. T	5. F
6. T	7. T	8. T	9. F	10. T

11. Laurie is friendly and Marcia is not happy.
12. Laurie is friendly and Marcia is not happy or John is grouchy.
13. If two triangles do not have the same area, then they are not congruent.
14.

p	q	$p \vee q$	$\sim(p \vee q)$	$\sim p$	$\sim q$	$\sim p \wedge \sim q$	$\sim(p \wedge q) \Leftrightarrow \sim p \wedge \sim q$
T	T	T	F	F	F	F	T
T	F	T	F	F	T	F	T
F	T	T	F	T	F	F	T
F	F	F	T	T	T	T	T

15. I passed the class.
16. It is sunny.
17. You can go in.
18. No dogs are cats.
19. All math classes are interesting.
20. No sunny days are rainy.

CHAPTER 12 TEST FORM 1

1. T
2. F
3. T
4. F
5. T
6. F
7. T
8. F
9. T
10. T
11. 15
12. $\left(3, \frac{15}{2}\right)$
13. $\frac{5}{2}$
14. $y = \frac{1}{2}x + \frac{1}{2}$
15. $y = \frac{7}{12}x + \frac{22}{3}$
16. $\frac{3}{2}$
17. $y = 4$
18. $-\frac{4}{3}$
19. $x = 8$
20. $x = 4,\ y = 1$

CHAPTER 12 TEST FORM 2

1. T
2. F
3. F
4. T
5. F
6. T
7. T
8. T
9. T
10. F
11. 5
12. $(-3, 5)$
13. -1
14. $-\frac{3}{2}$
15. $y = \frac{3}{2}x$
16. $y = x - 3$
17. $y = 2x - 2$
18. $-\frac{3}{2}$
19. $y = 3$
20. $x = 3,\ y = -7$

CHAPTER 12 TEST FORM 3

1. F
2. T
3. T
4. F
5. F
6. F
7. T
8. T
9. F
10. T
11. 13
12. $(-3, 7)$
13. $-\frac{7}{10}$
14. $-\frac{4}{7}$
15. $y = \frac{3}{4}x + 2$
16. $y = x - 3$
17. $y = 4x - 1$
18. $\frac{7}{6}$
19. $x = 1$
20. $x = 3,\ y = 1$

FINAL EXAM FORM 1

1.	F	21.	T	41.	T	61.	F	81.	10 in
2.	T	22.	T	42.	T	62.	T	82.	$\frac{32}{3}$ in
3.	T	23.	T	43.	T	63.	T	83.	$\frac{50}{3}$ in
4.	F	24.	F	44.	F	64.	F	84.	$\alpha°$
5.	T	25.	F	45.	F	65.	F	85.	$\frac{200}{3}$ sq in
6.	F	26.	T	46.	T	66.	T	86.	60°
7.	T	27.	F	47.	F	67.	F	87.	30°
8.	T	28.	F	48.	T	68.	F	88.	100°
9.	T	29.	F	49.	F	69.	T	89.	140°
10.	T	30.	T	50.	F	70.	F	90.	130°
11.	T	31.	T	51.	T	71.	F	91.	36°
12.	T	32.	T	52.	F	72.	T	92.	108°
13.	T	33.	F	53.	T	73.	T	93.	54°
14.	T	34.	F	54.	F	74.	T	94.	72°
15.	F	35.	F	55.	T	75.	T	95.	$\sqrt{a^2 + 25}$ in
16.	F	36.	T	56.	T	76.	180°	96.	130°
17.	T	37.	T	57.	T	77.	90°	97.	100°
18.	F	38.	F	58.	F	78.	5	98.	50°
19.	T	39.	T	59.	T	79.	$\frac{5}{2}$	99.	100°
20.	F	40.	T	60.	T	80.	5π	100.	25π sq in

FINAL EXAM
FORM 2

1.	T	21.	T	41.	T	61.	F	81.	18 cm
2.	T	22.	F	42.	F	62.	T	82.	27 cm
3.	F	23.	T	43.	T	63.	T	83.	36 cm
4.	F	24.	F	44.	F	64.	F	84.	30°
5.	F	25.	F	45.	F	65.	F	85.	$162\sqrt{3}$ sq cm
6.	T	26.	T	46.	F	66.	F	86.	90°
7.	F	27.	F	47.	T	67.	T	87.	5°
8.	F	28.	T	48.	T	68.	T	88.	115°
9.	T	29.	T	49.	T	69.	T	89.	140°
10.	T	30.	F	50.	F	70.	F	90.	120°
11.	T	31.	T	51.	T	71.	T	91.	36°
12.	T	32.	T	52.	F	72.	F	92.	108°
13.	F	33.	T	53.	F	73.	F	93.	54°
14.	F	34.	F	54.	F	74.	T	94.	72°
15.	T	35.	T	55.	T	75.	T	95.	$20a$ sq in
16.	T	36.	F	56.	T	76.	180°	96.	110°
17.	F	37.	T	57.	F	77.	90°	97.	140°
18.	T	38.	F	58.	T	78.	10	98.	70°
19.	T	39.	T	59.	T	79.	5	99.	140°
20.	T	40.	T	60.	T	80.	25π	100.	36π sq in

FINAL EXAM
FORM 3

1.	F	21.	T	41.	T	61.	T	81.	10 in
2.	F	22.	T	42.	T	62.	F	82.	8 in
3.	T	23.	F	43.	F	63.	F	83.	28 in
4.	T	24.	T	44.	F	64.	T	84.	$180^\circ - \alpha^\circ$
5.	T	25.	F	45.	T	65.	T	85.	$8\sqrt{21}$ sq in
6.	F	26.	T	46.	F	66.	F	86.	120°
7.	T	27.	T	47.	T	67.	T	87.	60°
8.	T	28.	F	48.	T	68.	T	88.	10
9.	F	29.	T	49.	F	69.	T	89.	5
10.	F	30.	T	50.	F	70.	F	90.	25π
11.	F	31.	T	51.	F	71.	F	91.	30°
12.	T	32.	T	52.	T	72.	F	92.	120°
13.	T	33.	F	53.	T	73.	T	93.	60°
14.	T	34.	F	54.	F	74.	F	94.	60°
15.	F	35.	F	55.	T	75.	T	95.	$15a$ sq cm
16.	T	36.	T	56.	F	76.	40°	96.	70°
17.	F	37.	F	57.	T	77.	64°	97.	110°
18.	F	38.	T	58.	F	78.	20°	98.	35°
19.	T	39.	F	59.	T	79.	44°	99.	10°
20.	T	40.	T	60.	T	80.	132°	100.	12π in

Exercise 1.5 (page 34)

16. Given: D is in the interior of $\angle ABC$.
P is in the interior of $\angle RST$.
$m(\angle ABC) = m(\angle RST)$.
$m(\angle ABD) = m(\angle RSP)$.

Prove: $m(\angle DBC) = m(\angle PST)$.

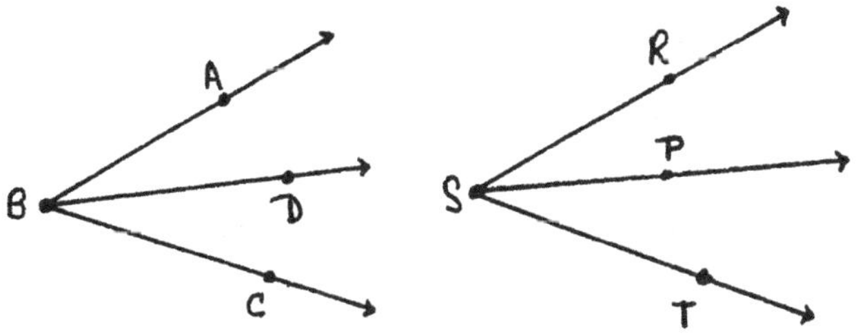

STATEMENTS	REASONS
1. D is in the interior of $\angle ABC$.	1. Given.
2. P is in the interior of $\angle RST$.	2. Given.
3. $m(\angle ABC) = m(\angle RST)$.	3. Given.
4. $m(\angle ABD) = m(\angle RSP)$.	4. Given.
5. $m(\angle ABC) - m(\angle ABD) = m(\angle RST) - m(\angle RSP)$.	5. If equal quantities are subtracted from equal quantities, the results are equal quantities.
6. $m(\angle DBC) = m(\angle PST)$.	6. Postulate 1.11.

18. Given: $\overrightarrow{BD}$ bisects $\angle ABC$.

Prove: $m(\angle ABD) = \frac{m(\angle ABC)}{2}$.

$m(\angle DBC) = \frac{m(\angle ABC)}{2}$.

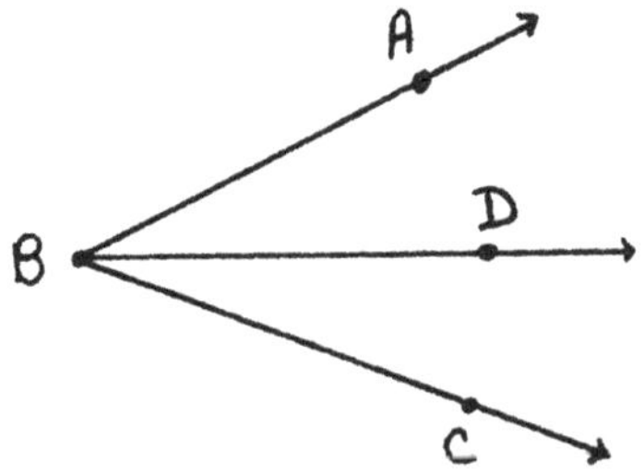

STATEMENTS	REASONS
1. $\overrightarrow{BD}$ bisects $\angle ABC$.	1. Given.
2. $\angle ABD \cong \angle DBC$.	2. An angle bisector divides an angle into two congruent angles.
3. $m(\angle ABD) = m(\angle DBC)$.	3. Congruent angles have equal measure.
4. $m(\angle ABC) = m(\angle ABD) + m(\angle DBC)$.	4. Postulate 1.11.
5. $m(\angle ABC) = m(\angle ABD) + m(\angle ABD)$.	5. Use postulate 1.9 to substitute $m(\angle ABD)$ for $m(\angle DBC)$.
6. $m(\angle ABC) = 2\,m(\angle ABD)$.	6. Use postulate 1.9 to substitute $2\,m(\angle ABD)$ for $m(\angle ABD) + m(\angle ABD)$.
7. $\frac{m(\angle ABC)}{2} = m(\angle ABD)$	7. If equal quantities are divided by nonzero equal quantities, the results are equal quantities.

The proof that $m(\angle DBC) = \frac{m(\angle ABC)}{2}$ is similar.

20. Given: $m(\overline{AB}) = m(\overline{CD})$.

Prove: $m(\overline{AC}) = m(\overline{BD})$.

STATEMENTS	REASONS
1. $m(\overline{AB}) = m(\overline{CD})$.	1. Given.
2. $m(\overline{BC}) = m(\overline{BC})$.	2. Reflexive law.
3. $m(\overline{AB}) + m(\overline{BC}) = m(\overline{BC}) + m(\overline{CD})$.	3. Equal quantities added to equal quantities are equal.
4. $m(\overline{AC}) = m(\overline{BD})$	4. Postulate 1.10.

22. Given: $m(\angle AOC) = m(\angle BOD)$.

Prove: $m(\angle AOB) = m(\angle COD)$.

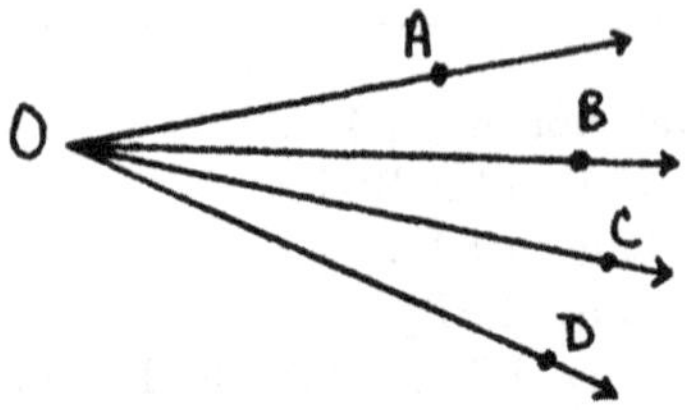

STATEMENTS	REASONS
1. $m(\angle AOC) = m(\angle BOD)$.	1. Given.
2. $m(\angle BOC) = m(\angle BOC)$.	2. Reflexive law.
3. $m(\angle AOC) - m(\angle BOC) = m(\angle BOD) - m(\angle BOC)$.	3. Equal quantities subtracted from equal quantities are equal.
4. $m(\angle AOB) = m(\angle COD)$.	4. Postulate 1.11.

Exercise 1.6 (page 41)

22. Given: $\angle C$ is complementary to $\angle A$.
$\angle D$ is complementary to $\angle B$.
$\angle A \cong \angle B$.

Prove: $\angle C \cong \angle D$.

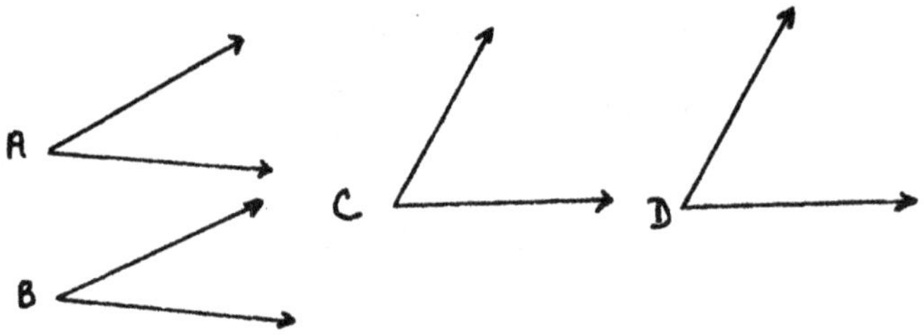

STATEMENTS	REASONS
1. $\angle C$ is complementary to $\angle A$.	1. Given.
2. $\angle D$ is complementary to $\angle B$.	2. Given.
3. $\angle A \cong \angle B$.	3. Given.
4. $m(\angle C) + m(\angle A) = 90^\circ$.	4. If two angles are complementary, then the sum of their measures is 90°.
5. $m(\angle D) + m(\angle B) = 90^\circ$.	5. If two angles are complementary, then the sum of their measures is 90°.
6. $m(\angle C) + m(\angle A) = m(\angle D) + m(\angle B)$.	6. Substitution.
7. $m(\angle A) = m(\angle B)$.	7. If two angles are congruent, they have equal measure.
8. $m(\angle C) = m(\angle D)$.	8. If equal quantities in step 7 are subtracted from quantities in step **6**, then the results are equal quantities.
9. $\angle C \cong \angle D$.	9. If two angles have equal measures, they are congruent.

24. Given: $\angle 1$ and $\angle 2$ are complementary.

Prove: $\angle 3$ and $\angle 4$ are complementary.

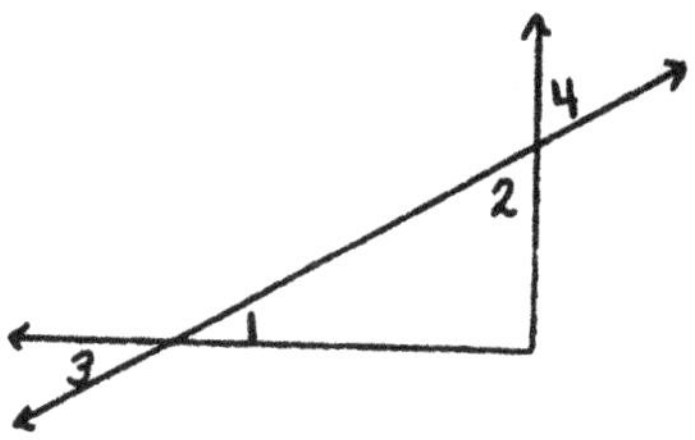

STATEMENTS	REASONS
1. $\angle 1$ and $\angle 2$ are complementary.	1. Given.
2. $m(\angle 1) + m(\angle 2) = 90°$.	2. If two angles are complementary, then the sum of their measures is 90°.
3. $\angle 3$ and $\angle 1$ are vertical angles.	3. A pair of nonadjacent angles formed by two intersecting lines is a pair of vertical angles.
4. $\angle 3 \cong \angle 1$.	4. Vertical angles are congruent.
5. $m(\angle 3) = m(\angle 1)$.	5. Congruent angles have equal measure.
6. $\angle 2$ and $\angle 4$ are vertical angles.	6. A pair of nonadjacent angles formed by two intersecting lines is a pair of vertical angles.
7. $\angle 2 \cong \angle 4$.	7. Vertical angles are congruent.
8. $m(\angle 2) = m(\angle 4)$.	8. Congruent angles have equal measure.
9. $m(\angle 3) + m(\angle 4) = 90°$.	9. Substitution.
10. $\angle 3$ and $\angle 4$ are complementary.	10. If the sum of the measures of two angles is 90°, the angles are complementary.

Chapter One Review Exercises (page 43)

38. Given: B is the midpoint of $\overline{AC}$.

D is the midpoint of $\overline{CE}$.

$\overline{AC} \cong \overline{CE}$.

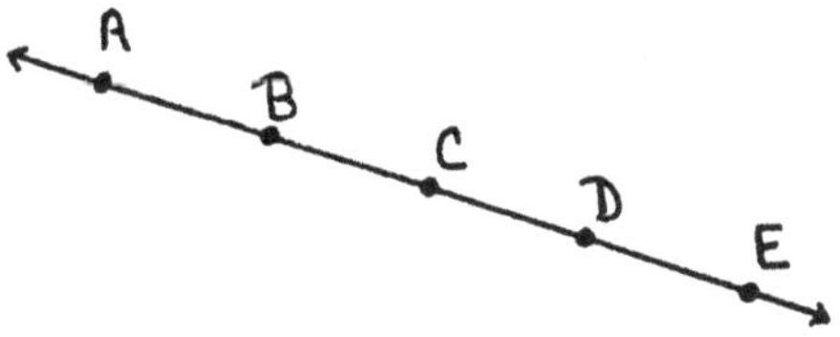

Prove: $\overline{AB} \cong \overline{DE}$.

STATEMENTS	REASONS
1. B is the midpoint of $\overline{AB}$.	1. Given.
2. D is the midpoint of $\overline{CE}$.	2. Given.
3. $\overline{AC} \cong \overline{CE}$.	3. Given.
4. $m(\overline{AC}) = m(\overline{CE})$.	4. Congruent line segments have equal measure.
5. $m(\overline{AB}) = m(\overline{DE})$.	5. Theorem 1.6.
6. $\overline{AB} \cong \overline{DE}$.	6. If line segments have equal measure then they are congruent.

Exercise 2.1 (page 53)

14. Given: $\overline{AB} \cong \overline{CD}$.

$\angle 1 \cong \angle 2$.

Prove: $\triangle ABD \cong \triangle CDB$.

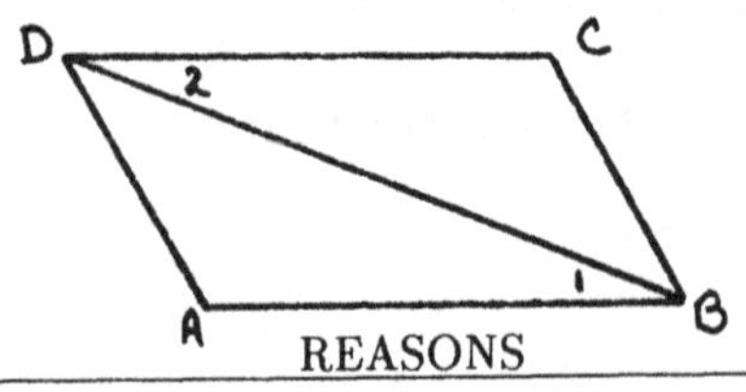

STATEMENTS	REASONS
1. $\overline{AB} \cong \overline{CD}$.	1. Given.
2. $\angle 1 \cong \angle 2$.	2. Given.
3. $\overline{BD} \cong \overline{DB}$.	3. A quantity is congruent to itself. (Reflexive law)
4. $\triangle ABD \cong \triangle CDB$.	4. SAS $\cong$ SAS.

16. Given: $\overline{AB} \cong \overline{DC}$.

$\overline{AC} \cong \overline{DB}$.

Prove: $\triangle ABC \cong \triangle DCB$.

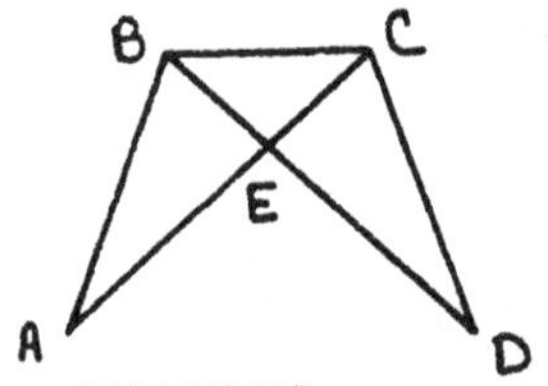

STATEMENTS	REASONS
1. $\overline{AB} \cong \overline{DC}$.	1. Given.
2. $\overline{AC} \cong \overline{DB}$.	2. Given.
3. $\overline{BC} \cong \overline{CB}$.	3. A quantity is congruent to itself. (Reflexive law)
4. $\triangle ABC \cong \triangle DCB$.	4. SSS $\cong$ SSS.

18. Given: $\overline{AD} \cong \overline{BD}$.

$\angle 1 \cong \angle 2$.

Prove: $\triangle ADC \cong \triangle BDC$.

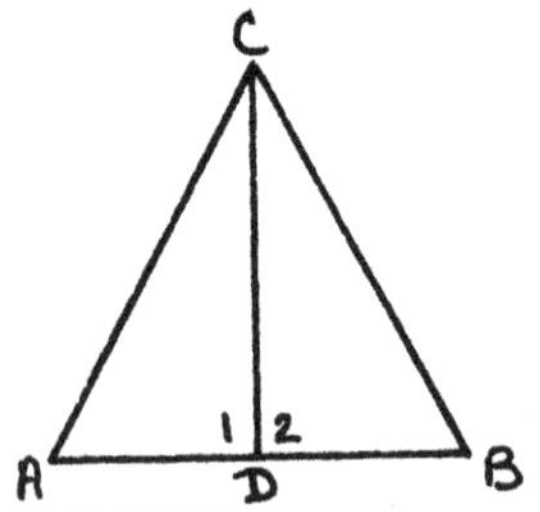

STATEMENTS	REASONS
1. $\overline{AD} \cong \overline{BD}$.	1. Given.
2. $\angle 1 \cong \angle 2$.	2. Given.
3. $\overline{DC} \cong \overline{DC}$.	3. A quantity is congruent to itself. (Reflexive law)
4. $\triangle ADC \cong \triangle BDC$.	4. SAS $\cong$ SAS.

20. Given: $\overline{AE}$ and $\overline{BD}$ are line segments intersecting at point C.

$\overline{AC} \cong \overline{EC}$.

$\overline{BC} \cong \overline{DC}$.

Prove: $\triangle ABC \cong \triangle EDC$.

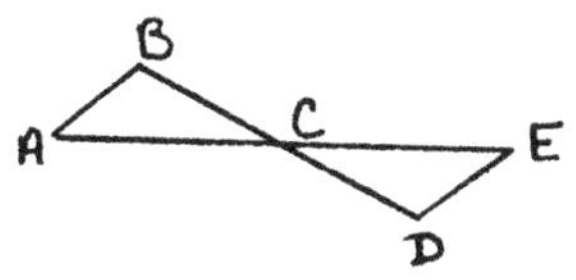

STATEMENTS	REASONS
1. $\overline{AE}$ and $\overline{BD}$ are line segments intersecting at point C.	1. Given.
2. $\overline{AC} \cong \overline{EC}$.	2. Given.
3. $\overline{BC} \cong \overline{DC}$.	3. Given.
4. $\angle BCA$ and $\angle DCE$ are vertical angles.	4. A pair of nonadjacent angles formed by two intersecting lines is a pair of vertical angles.
5. $\angle BCA \cong \angle DCE$.	5. Vertical angles are congruent.
6. $\triangle ABC \cong \triangle EDC$.	6. SAS $\cong$ SAS.

22. Given: $\triangle ADE$.

$\angle A \cong \angle D$.

$\overline{AE} \cong \overline{DE}$.

$\overline{AC} \cong \overline{DB}$.

Prove: $\triangle ABE \cong \triangle DCE$.

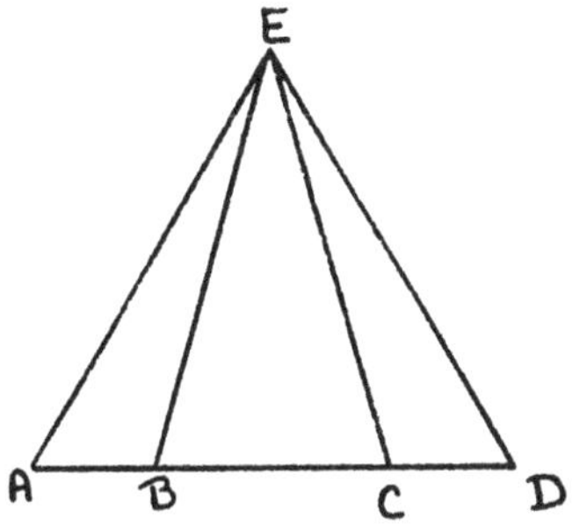

STATEMENTS	REASONS
1. $\triangle ADE$.	1. Given.
2. $\angle A \cong \angle D$.	2. Given.
3. $\overline{AE} \cong \overline{DE}$.	3. Given.
4. $\overline{AC} \cong \overline{DB}$.	4. Given.
5. $m(\overline{AC}) = m(\overline{DB})$.	5. Congruent line segments have equal measure.
6. $m(\overline{BC}) = m(\overline{CB})$.	6. Reflexive law.
7. $m(\overline{AC}) - m(\overline{BC}) = m(\overline{DB}) - m(\overline{CB})$.	7. Equal quantities subtracted from equal quantities are equal.
8. $m(\overline{AC}) - m(\overline{BC}) = m(\overline{AB})$.	8. Postulate 1.10.
9. $m(\overline{DB}) - m(\overline{CB}) = m(\overline{DC})$.	9. Postulate 1.10.
10. $m(\overline{AB}) = m(\overline{DC})$.	10. Substitution.
11. $\overline{AB} \cong \overline{DC}$.	11. Line segments with equal measure are congruent.
12. $\triangle ABE \cong \triangle DCE$.	12. SAS $\cong$ SAS.

24. Given: $\angle A \cong \angle B$.
$\overline{AC} \cong \overline{BC}$.

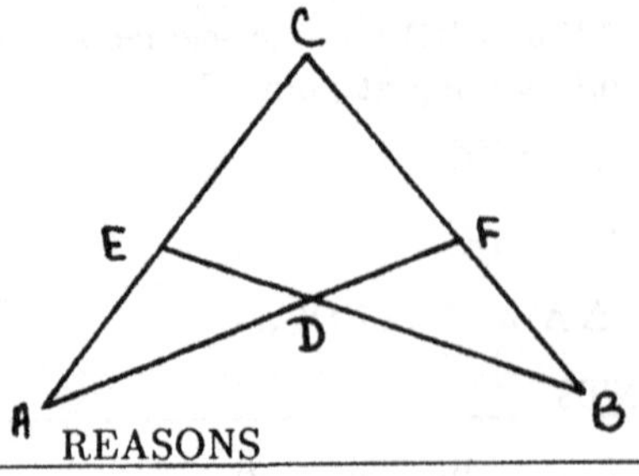

Prove: $\triangle AFC \cong \triangle BEC$.

STATEMENTS	REASONS
1. $\angle A \cong \angle B$.	1. Given.
2. $\overline{AC} \cong \overline{BC}$.	2. Given.
3. $\angle C \cong \angle C$.	3. Reflexive law.
4. $\triangle AFC \cong \triangle BEC$.	4. ASA $\cong$ ASA.

26. Given: $\overline{AD} \cong \overline{AC}$.
$\overline{AB} \cong \overline{AE}$.

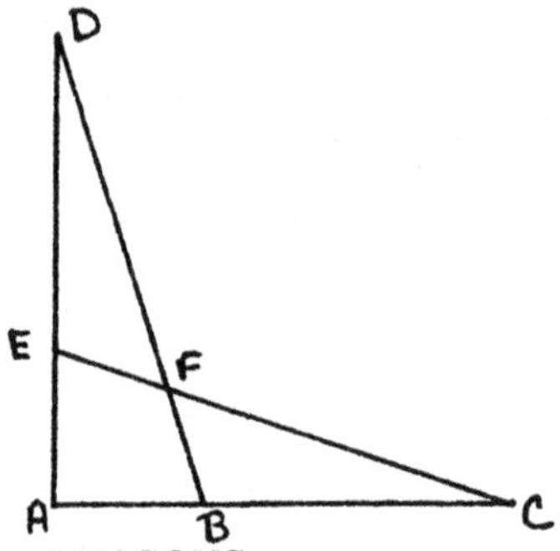

Prove: $\triangle ADB \cong \triangle ACE$.

STATEMENTS	REASONS
1. $\overline{AD} \cong \overline{AC}$.	1. Given.
2. $\overline{AB} \cong \overline{AE}$.	2. Given.
3. $\angle A \cong \angle A$.	3. Reflexive law.
4. $\triangle ADB \cong \triangle ACE$.	4. SAS $\cong$ SAS.

Exercise 2.2 (page 60)

2. Given: $\overline{AB} \cong \overline{CD}$.
$\overline{BC} \cong \overline{DA}$.
Prove: $\angle 1 \cong \angle 2$.

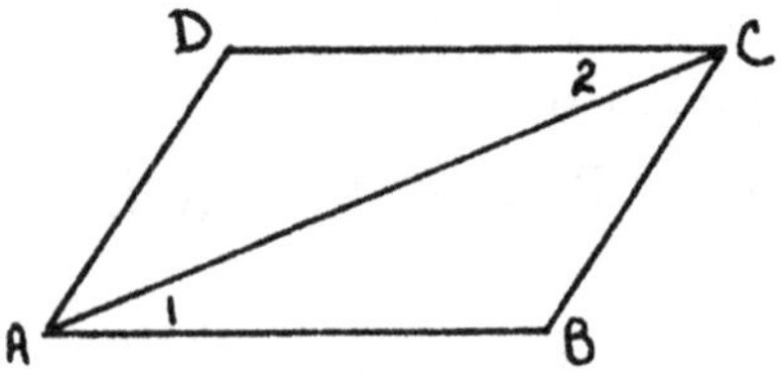

STATEMENTS	REASONS
1. $\overline{AB} \cong \overline{CD}$.	1. Given.
2. $\overline{BC} \cong \overline{DA}$.	2. Given.
3. $\overline{AC} \cong \overline{AC}$.	3. A quantity is congruent to itself. (Reflexive law)
4. $\triangle ABC \cong \triangle CDA$.	4. SSS $\cong$ SSS.
5. $\angle 1 \cong \angle 2$.	5. cpctc.

4. Given: $\overline{BF}$ and $\overline{DE}$ intersecting at point C.
$\overline{AB} \cong \overline{AD}$.
$\angle B \cong \angle D$.

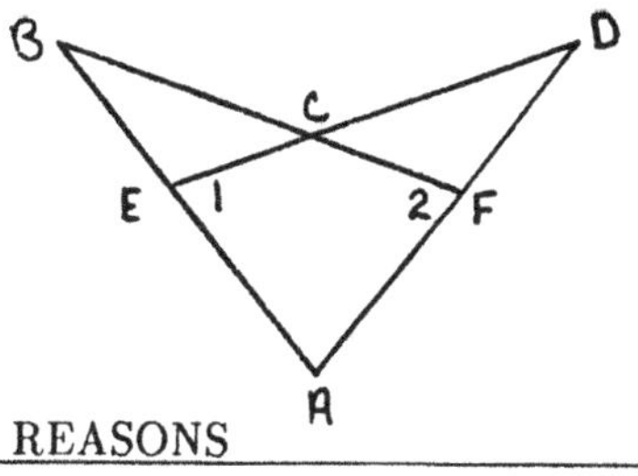

Prove: $\angle 2 \cong \angle 1$.

STATEMENTS	REASONS
1. $\overline{BF}$ and $\overline{DE}$ intersecting at point C.	1. Given.
2. $\overline{AB} \cong \overline{AD}$.	2. Given.
3. $\angle B \cong \angle D$.	3. Given.
4. $\angle A \cong \angle A$.	4. An angle is congruent to itself. (Reflexive law)
5. $\triangle ABF \cong \triangle ADE$.	5. ASA $\cong$ ASA.
6. $\angle 2 \cong \angle 1$.	6. cpctc.

6. Given: $\overline{DB} \cong \overline{EA}$.
$\overline{AD} \cong \overline{BE}$.

Prove: $\angle DAB \cong \angle EBA$.

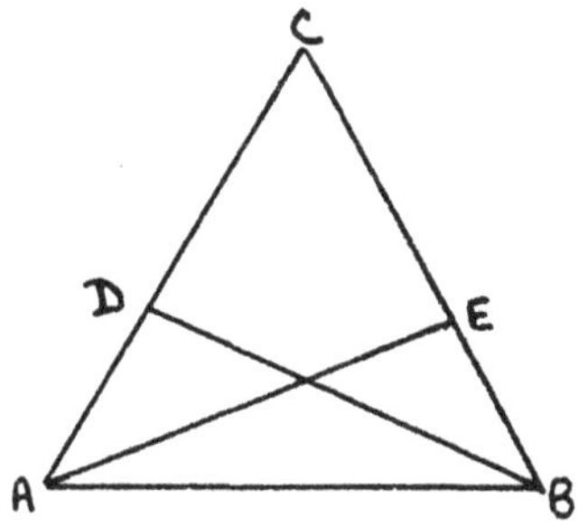

STATEMENTS	REASONS
1. $\overline{DB} \cong \overline{EA}$.	1. Given.
2. $\overline{AD} \cong \overline{BE}$.	2. Given.
3. $\overline{AB} \cong \overline{BA}$.	3. Reflexive law.
4. $\triangle ADB \cong \triangle BEA$.	4. SSS $\cong$ SSS.
5. $\angle DAB \cong \angle EBA$.	5. cpctc.

8. Given: $\triangle ABC$.
$\overline{DC} \cong \overline{EC}$.
$\angle 1 \cong \angle 2$.

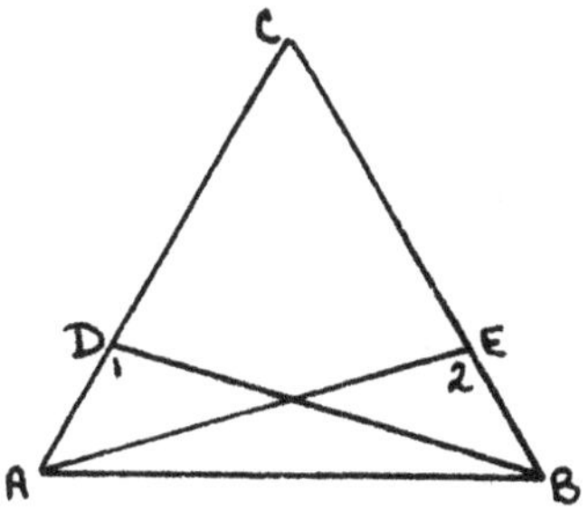

Prove: $\overline{AC} \cong \overline{BC}$.

STATEMENTS	REASONS
1. $\triangle ABC$.	1. Given.
2. $\overline{DC} \cong \overline{EC}$.	2. Given.

Continued on the next page.

STATEMENTS	REASONS
3. $\angle 1 \cong \angle 2$.	3. Given.
4. $\angle C \cong \angle C$.	4. Reflexive law.
5. $\angle 1$ and $\angle CDB$ are supplmentary, $\angle 2$ and $\angle CEA$ are supplementary.	5. Adjacent angles whose noncommon sides form a line are supplementary.
6. $\angle CDB \cong \angle CEA$.	6. Supplements of congruent angles are congruent.
7. $\triangle CDB \cong \triangle CEA$.	7. ASA $\cong$ ASA.
8. $\overline{AC} \cong \overline{BC}$.	8. cpctc.

10. Given: $\angle PAQ$.

Prove: $\overline{AD}$ bisects $\angle A$.

Construction: $\overline{AB} \cong \overline{AC}$.

$\overline{BD} \cong \overline{CD}$.

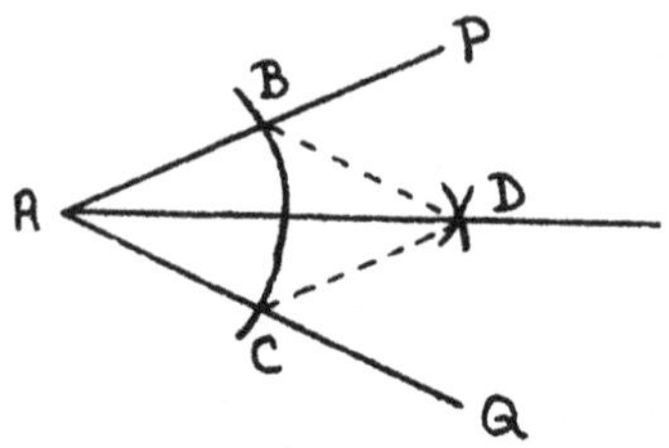

STATEMENTS	REASONS
1. $\angle PAQ$.	1. Given.
2. $\overline{AB} \cong \overline{AC}$.	2. By construction.
3. $\overline{BD} \cong \overline{CD}$.	3. By construction.
4. $\overline{AD} \cong \overline{AD}$.	4. Reflexive law.
5. $\triangle ABD \cong \triangle ACD$.	5. SSS $\cong$ SSS.
6. $\angle BAD \cong \angle CAD$.	6. cpctc.
7. $\overline{AD}$ bisects $\angle A$.	7. If a segment divides an angle into two congruent angles, it bisects the angle.

12. Given: Line l and point A not on line l.

Prove: $\overline{AD} \perp \overline{BC}$.

Construction: $\overline{AB} \cong \overline{AC}$.

$\overline{BD} \cong \overline{CD}$.

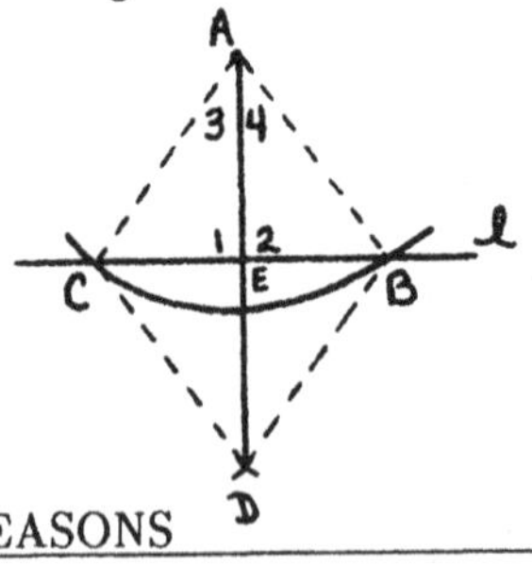

STATEMENTS	REASONS
1. Line l and point A not on line l.	1. Given.
2. $\overline{AB} \cong \overline{AC}$.	2. By construction.
3. $\overline{BD} \cong \overline{CD}$.	3. By construction.

Continued on the next page.

STATEMENTS	REASONS
4. $\overline{AD} \cong \overline{AD}$.	4. Reflexive law.
5. $\triangle ABD \cong \triangle ACD$.	5. SSS ≅ SSS.
6. $\angle 3 \cong \angle 4$.	6. cpctc.
7. $\overline{AE} \cong \overline{AE}$.	7. Reflexive law.
8. $\triangle ABE \cong \triangle ACE$.	8. SAS ≅ SAS.
9. $\angle 2 \cong \angle 1$.	9. cpctc.
10. $\angle 1$ and $\angle 2$ are adjacent angles.	10. Two angles with a common vertex and a common side between them are adjacent angles.
11. $\overline{AD} \perp \overline{BC}$.	11. If two lines meet and form congruent adjacent angles, the lines are perpendicular.

14. Given: $\overline{DC} \cong \overline{AB}$.

$\angle CDA \cong \angle BAD$.

Prove: $\overline{AC} \cong \overline{DB}$.

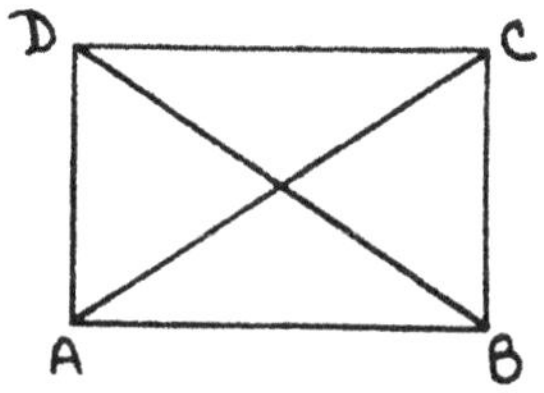

STATEMENTS	REASONS
1. $\overline{DC} \cong \overline{AB}$.	1. Given.
2. $\angle CDA \cong \angle BAD$.	2. Given.
3. $\overline{DA} \cong \overline{AD}$.	3. A quantity is congruent to itself. (Reflexive law)
4. $\triangle CDA \cong \triangle BAD$.	4. SAS ≅ SAS.
5. $\overline{AC} \cong \overline{DB}$.	5. cpctc.

Exercise 2.3 (page 68)

8. Given: $\triangle ABC$ is an isosceles triangle with $\overline{AB} \cong \overline{AC}$.

$\overline{AM}$ is a median drawn to $\overline{BC}$.

Prove: $\overline{AM} \perp \overline{BC}$.

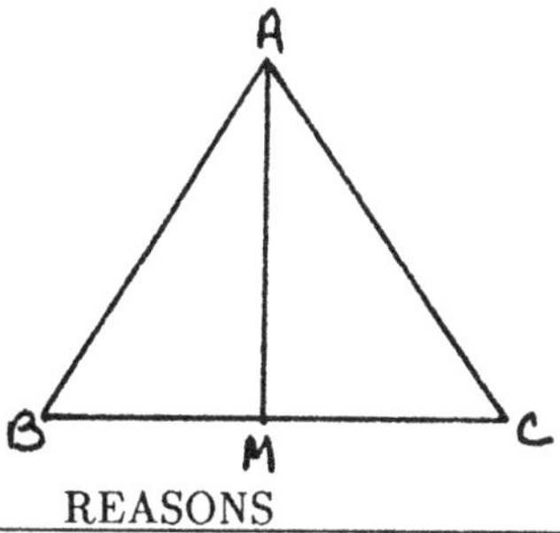

STATEMENTS	REASONS
1. $\triangle ABC$ is an isosceles triangle with $\overline{AB} \cong \overline{AC}$.	1. Given.
2. $\overline{AM}$ is a median drawn to $\overline{BC}$.	2. Given.

Continued on the next page.

STATEMENTS	REASONS
3. $\overline{BM} \cong \overline{CM}$.	3. Median $\overline{AM}$ bisects $\overline{BC}$.
4. $\overline{AM} \cong \overline{AM}$.	4. Reflexive law.
5. $\triangle ABM \cong \triangle ACM$.	5. SSS $\cong$ SSS.
6. $\angle 1 \cong \angle 2$.	6. cpctc.
7. $\angle 1$ and $\angle 2$ are adjacent angles.	7. Two angles with a common vertex and a common side between them are adjacent angles.
8. $\overline{AM} \perp \overline{BC}$.	8. If two lines meet and form congruent adjacent angles, the lines are perpendicular.

10. Given: $\angle 1 \cong \angle 2$.

$\overline{AC} \cong \overline{BC}$.

Prove: $\angle 3 \cong \angle 4$.

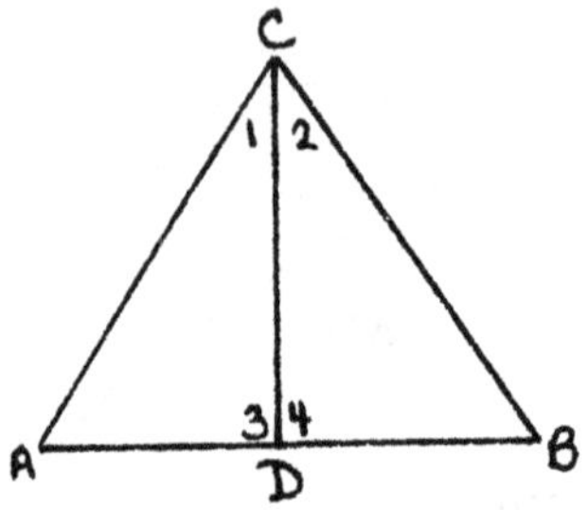

STATEMENTS	REASONS
1. $\angle 1 \cong \angle 2$.	1. Given.
2. $\overline{AC} \cong \overline{BC}$.	2. Given.
3. $\overline{CD} \cong \overline{CD}$.	3. Reflexive law.
4. $\triangle ACD \cong \triangle BCD$.	4. SAS $\cong$ SAS.
5. $\angle 3 \cong \angle 4$.	5. cpctc.

12. Given: $\overline{AC} \cong \overline{BC}$.

$\overline{AD} \cong \overline{BD}$.

Prove: $\angle CAD \cong \angle CBD$.

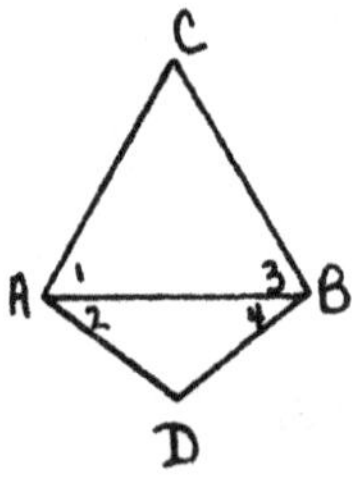

STATEMENTS	REASONS
1. $\overline{AC} \cong \overline{BC}$.	1. Given.
2. $\overline{AD} \cong \overline{BD}$.	2. Given.
3. $\angle 1 \cong \angle 3$.	3. Theorem 2.2.
4. $\angle 2 \cong \angle 4$.	4. Theorem 2.2.
5. $m(\angle 1) = m(\angle 3)$.	5. Congruent angles have equal measure.

Continued on the next page.

STATEMENTS	REASONS
6. $m(\angle 2) = m(\angle 4)$.	6. Congruent angles have equal measure.
7. $m(\angle 1) + m(\angle 2) = m(\angle 3) + m(\angle 4)$.	7. Equal quantities added to equal quantities are equal.
8. $m(\angle 1) + m(\angle 2) = m(\angle CAD)$.	8. Postulate 1.11.
9. $m(\angle 3) + m(\angle 4) = m(\angle CBD)$.	9. Postulate 1.11.
10. $m(\angle CAD) = m(\angle CBD)$.	10. Substitution.
11. $\angle CAD \cong \angle CBD$.	11. Angles with equal measure are congruent.

14. Given: $\triangle ABC$ is an isosceles triangle with $\overline{AC} \cong \overline{BC}$. $\overline{CE}$ bisects $\overline{AB}$.

Prove: $\triangle ABD$ is isosceles.

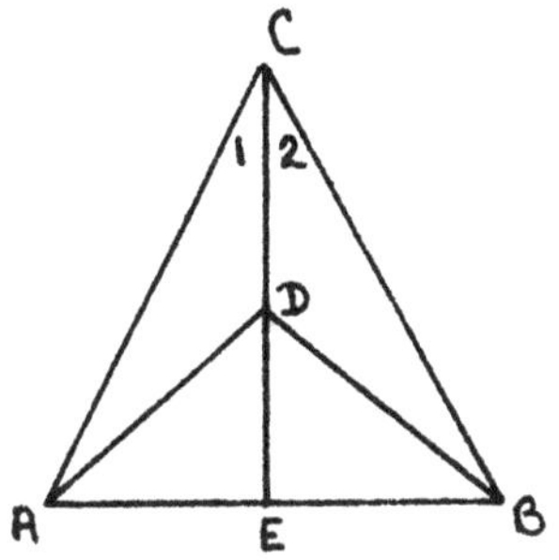

STATEMENTS	REASONS
1. $\triangle ABC$ is an isosceles triangle with $\overline{AC} \cong \overline{BC}$.	1. Given.
2. $\overline{CE}$ bisects $\overline{AB}$.	2. Given.
3. $\overline{AE} \cong \overline{BE}$.	3. A bisector divides a segment into two congruent segments.
4. $\overline{CE} \cong \overline{CE}$.	4. Reflexive law.
5. $\triangle AEC \cong \triangle BEC$.	5. SSS $\cong$ SSS.
6. $\angle 1 \cong \angle 2$.	6. cpctc.
7. $\overline{CD} \cong \overline{CD}$.	7. Reflexive law.
8. $\triangle ACD \cong \triangle BCD$.	8. SAS $\cong$ SAS.
9. $\overline{AD} \cong \overline{BD}$.	9. cpctc.
10. $\triangle ABD$ is isosceles.	10. Because $\triangle ABD$ has two congruent sides, it is isosceles.

16. Given: Isosceles $\triangle ABC$ and $\triangle ABD$ with common base $\overline{AB}$.

Prove: $\angle 1 \cong \angle 2$.

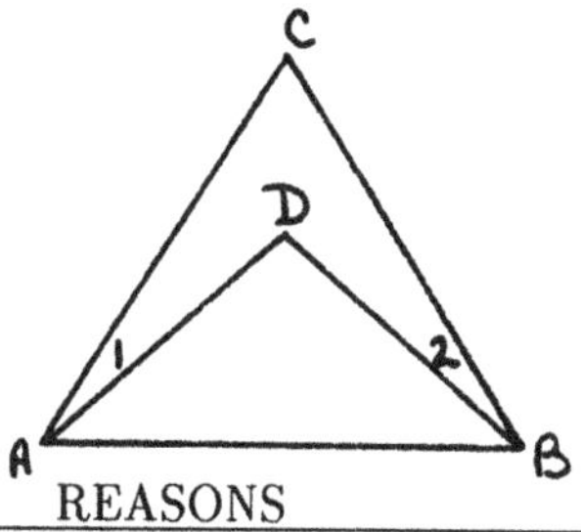

STATEMENTS	REASONS
1. Isosceles $\triangle ABC$ and $\triangle ABD$ with base $\overline{AB}$.	1. Given.

Continued on the next page.

STATEMENTS	REASONS
2. $\overline{AC} \cong \overline{BC}$.	2. An isosceles triangle has two congruent sides.
3. $\angle CAB \cong \angle CBA$.	3. Base angles of an isosceles triangle are congruent.
4. $\overline{AD} \cong \overline{BD}$.	4. An isosceles triangle has two congruent sides.
5. $\angle DAB \cong \angle DBA$.	5. Base angles of an isosceles triangle are congruent.
6. $m(\angle CAB) = m(\angle CBA)$. $m(\angle DAB) = m(\angle DBA)$.	6. Congruent angles have equal measure.
7. $m(\angle CAB) - m(\angle DAB) = m(\angle CBA) - m(\angle DBA)$.	7. Equal quantities subtracted from equal quantities are equal.
8. $m(\angle CAB) - m(\angle DAB) = m(\angle 1)$.	8. Postulate 1.11.
9. $m(\angle CBA) - m(\angle DBA) = m(\angle 2)$.	9. Postulate 1.11.
10. $m(\angle 1) = m(\angle 2)$.	10. Substitution.
11. $\angle 1 \cong \angle 2$.	11. Angles with equal measure are congruent.

Exercise 2.4 (page 73)

10. Given: $\triangle ABE$.

$\angle 1 \cong \angle 2$.

$\overline{AC} \cong \overline{BD}$.

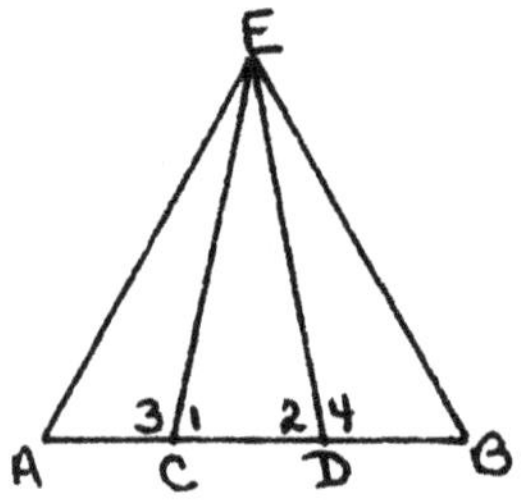

Prove: $\angle A \cong \angle B$.

STATEMENTS	REASONS
1. $\triangle ABE$.	1. Given.
2. $\angle 1 \cong \angle 2$.	2. Given.
3. $\overline{AC} \cong \overline{BD}$.	3. Given.
4. $\overline{CE} \cong \overline{DE}$.	4. If a triangle has two congruent angles, then the sides opposite those angles are congruent.
5. $\angle 1$ and $\angle 3$ are supplementary.	5. Adjacent angles whose noncommon sides form a line are supplementary.

Continued on the next page.

STATEMENTS	REASONS
6. $\angle 2$ and $\angle 4$ are supplementary.	6. Adjacent angles whose noncommon sides form a line are supplementary.
7. $\angle 3 \cong \angle 4$.	7. Supplements of congruent angles are congruent.
8. $\triangle ACE \cong \triangle BDE$.	8. SAS $\cong$ SAS.
9. $\angle A \cong \angle B$.	9. cpctc.

12. Given: $\angle 1 \cong \angle 2$.
$\angle 3 \cong \angle 4$.

Prove: $\angle CAD \cong \angle CBD$.

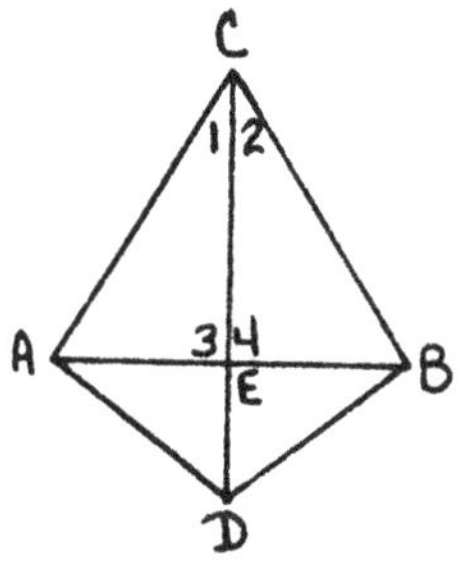

STATEMENTS	REASONS
1. $\angle 1 \cong \angle 2$.	1. Given.
2. $\angle 3 \cong \angle 4$.	2. Given.
3. $\overline{CE} \cong \overline{CE}$.	3. Reflexive law.
4. $\triangle ACE \cong \triangle BEC$.	4. ASA $\cong$ ASA.
5. $\overline{AC} \cong \overline{BC}$.	5. cpctc.
6. $\overline{CD} \cong \overline{CD}$.	6. Reflexive law.
7. $\triangle ACD \cong \triangle BCD$.	7. SAS $\cong$ SAS.
8. $\angle CAD \cong \angle CBD$.	8. cpctc.

14. Given: $\angle 1 \cong \angle 2$.
$\angle 3 \cong \angle 4$.

Prove: $\angle ADC \cong \angle BEC$.

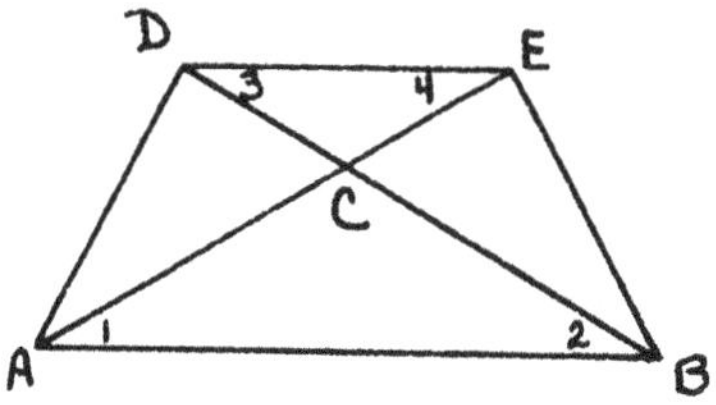

STATEMENTS	REASONS
1. $\angle 1 \cong \angle 2$.	1. Given.
2. $\angle 3 \cong \angle 4$.	2. Given.
3. $\overline{AC} \cong \overline{BC}$; $\overline{DC} \cong \overline{EC}$.	3. If a triangle has two congruent angles, then the sides opposite those angles are congruent.

Continued on the next page.

STATEMENTS	REASONS
4. ∠DCA and ∠ECB are vertical angles.	4. A pair of nonadjacent angles formed by two intersecting lines is a pair of vertical angles.
5. ∠DCA ≅ ∠ECB.	5. Vertical angles are congruent.
6. △ADC ≅ △BEC.	6. SAS ≅ SAS.
7. ∠ADC ≅ ∠BEC.	7. cpctc.

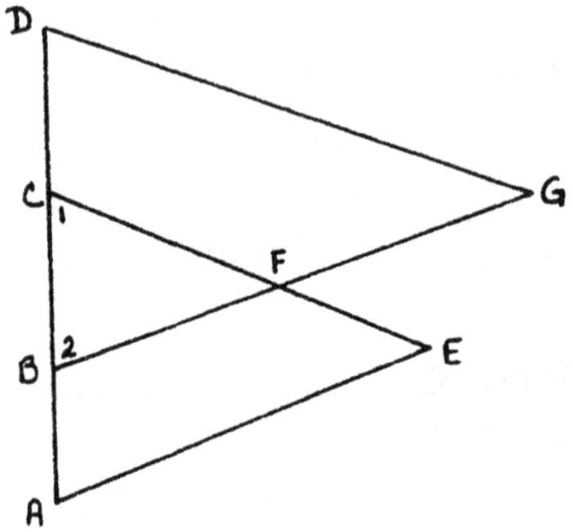

16. Given: Points A, B, C, and D lie on a line.

$\overline{CE} \cong \overline{AE}$.

∠2 ≅ ∠A.

Prove: △BCF is isosceles.

STATEMENTS	REASONS
1. Points A, B, C, and D lie on a line.	1. Given.
2. $\overline{CE} \cong \overline{AE}$.	2. Given.
3. ∠2 ≅ ∠A.	3. Given.
4. ∠1 ≅ ∠A.	4. If a triangle has two congruent sides, then the angles opposite those sides are congruent.
5. ∠1 ≅ ∠2.	5. Substitution.
6. $\overline{CF} \cong \overline{BF}$.	6. If a triangle has two congruent angles, then the sides opposite those angles are congruent.
7. △BCF is isosceles.	7. If a triangle has two congruent sides, it is isosceles.

Exercise 2.5 (page 78)

4. Given: $\overline{ED}$ and $\overline{AC}$ intersect at point B.

   $\overline{AE} \perp \overline{AC}$.

   $\overline{CD} \perp \overline{AC}$.

   B is the midpoint of $\overline{ED}$.

   $\overline{AE} \cong \overline{CD}$.

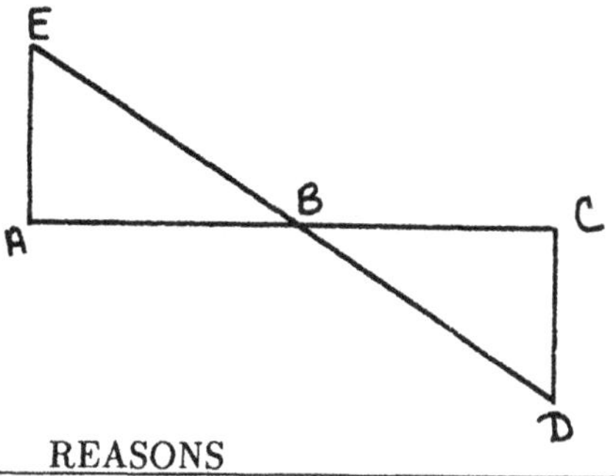

Prove: B is the midpoint of $\overline{AC}$.

STATEMENTS	REASONS
1. $\overline{ED}$ and $\overline{AC}$ intersect at point B.	1. Given.
2. $\overline{AE} \perp \overline{AC}$; $\overline{CD} \perp \overline{AC}$.	2. Given.
3. B is the midpoint of $\overline{ED}$.	3. Given.
4. $\overline{AE} \cong \overline{CD}$.	4. Given.
5. $\overline{EB} \cong \overline{DB}$.	5. A midpoint divides a line segment in two congruent segments.
6. $\angle$EAB and $\angle$DCB are right angles.	6. Perpendicular lines meet and form right angles.
7. $\triangle$ABE and $\triangle$CBD are right triangles.	7. If a triangle contains a right angle, then it is a right triangle.
8. $\triangle ABE \cong \triangle CBD$.	8. HL $\cong$ HL
9. $\overline{AB} \cong \overline{CB}$.	9. cpctc.
10. B is the midpoint of $\overline{AC}$.	10. If a point divides a line segment in two congruent segments, it is the midpoint of the line segment.

6. Given: $\overline{DB}$ and $\overline{AC}$ intersect at point O.

   $m(\angle 1) = m(\angle 2) = 90°$.

   $\overline{CD} \cong \overline{CB}$.

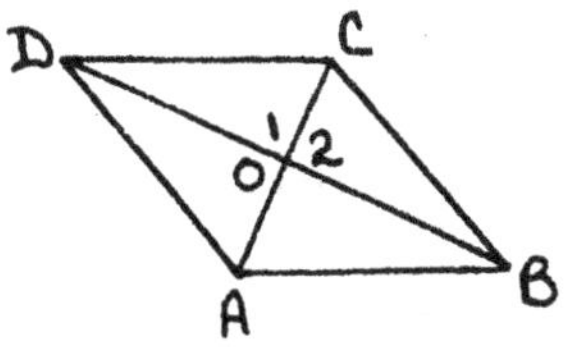

Prove: $\overline{OD} \cong \overline{OB}$.

The proof is on the next page.

STATEMENTS	REASONS
1. $\overline{DB}$ and $\overline{AC}$ intersect at point O.	1. Given.
2. $m(\angle 1) = m(\angle 2) = 90°$.	2. Given.
3. $\overline{CD} \cong \overline{CB}$.	3. Given.
4. $\angle 1$ and $\angle 2$ are right angles.	4. An angle with measure of 90° is a right angle.
5. $\triangle ODC$ and $\triangle OBC$ are right triangles.	5. If a triangle contains a right angle, then it is a right triangle.
6. $\overline{OC} \cong \overline{OC}$.	6. Reflexive law.
7. $\triangle ODC \cong \triangle OBC$.	7. HL $\cong$ HL.
8. $\overline{OD} \cong \overline{OB}$.	8. cpctc.

8. Given: $\triangle ABC$ is isosceles with base $\overline{AB}$.

$m(\overline{AD}) + m(\overline{DF}) = m(\overline{DC})$.

$\overline{BF} \perp \overline{CD}$.

$\overline{AD} \perp \overline{DC}$.

Prove: $\angle 3 \cong \angle 4$.

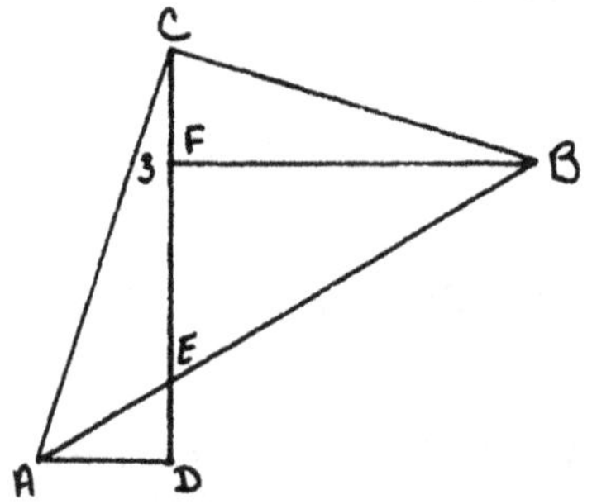

STATEMENTS	REASONS
1. $\triangle ABC$ is isosceles with base $\overline{AB}$.	1. Given.
2. $m(\overline{AD}) + m(\overline{DF}) = m(\overline{DC})$.	2. Given.
3. $\overline{BF} \perp \overline{CD}$; $\overline{AD} \perp \overline{DC}$.	3. Given.
4. $\overline{AC} \cong \overline{BC}$.	4. An isosceles triangle has two congruent sides.
5. $m(\overline{AD}) + m(\overline{DF}) - m(\overline{DF}) = m(\overline{DC}) - m(\overline{DF})$.	5. Equal quantities subtracted from equal quantities are equal.
6. $m(\overline{DC}) - m(\overline{DF}) = m(\overline{FC})$.	6. Postulate 1.10.
7. $m(\overline{AD}) = m(\overline{FC})$.	7. Substitution.
8. $\overline{AD} \cong \overline{FC}$.	8. Line segments with equal measure are congruent.
9. $\angle ADC$ and $\angle CFB$ are right angles.	9. Perpendicular lines meet and form right angles.
10. $\triangle ADC$ and $\triangle CFB$ are right triangles.	10. If a triangle contains a right angle, then it is a right triangle.
11. $\triangle ADC \cong \triangle CFB$.	11. HL $\cong$ HL
12. $\angle 3 \cong \angle 4$.	12. cpctc.

Chapter Two Review Exercises (page 79)

16. Given: $\overline{DE}$ and $\overline{AB}$ intersecting at point C.

$\overline{AC} \cong \overline{BC}$.

$\angle A \cong \angle B$.

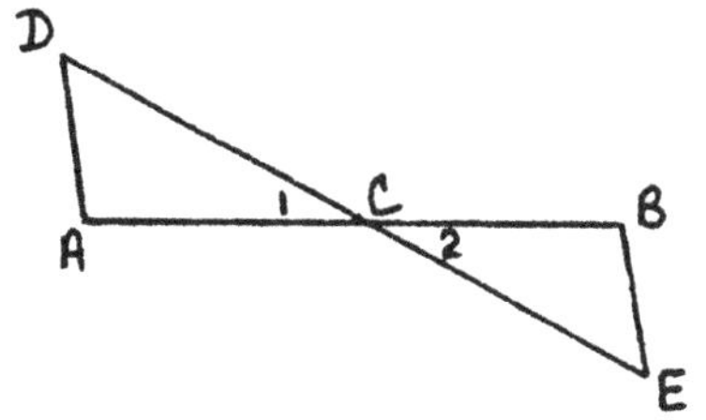

Prove: $\angle D \cong \angle E$.

STATEMENTS	REASONS
1. $\overline{DE}$ and $\overline{AB}$ intersecting at point C.	1. Given.
2. $\overline{AC} \cong \overline{BC}$.	2. Given.
3. $\angle A \cong \angle B$.	3. Given.
4. $\angle 1$ and $\angle 2$ are vertical angles.	4. Nonadjacent angles formed by intersecting lines segments are vertical angles.
5. $\angle 2 \cong \angle 1$.	5. Vertical angles are congruent.
6. $\triangle ACD \cong \triangle BCE$.	6. ASA ≅ ASA.
7. $\angle D \cong \angle E$.	7. cpctc.

26. Given: $\overline{BD} \cong \overline{BC}$.

$\angle 3 \cong \angle 2$.

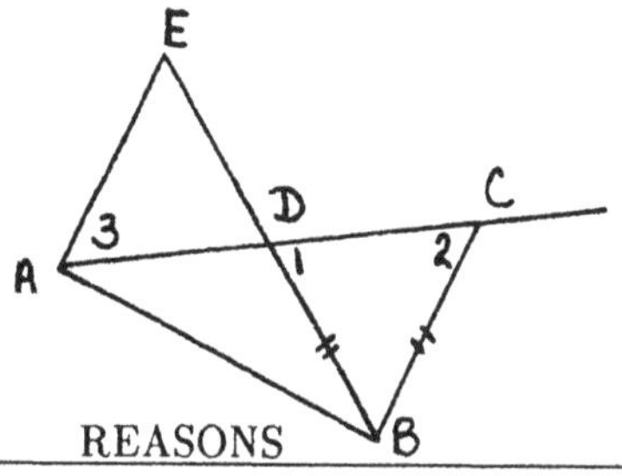

Prove: $\overline{AE} \cong \overline{DE}$.

STATEMENTS	REASONS
1. $\overline{BD} \cong \overline{BC}$.	1. Given.
2. $\angle 3 \cong \angle 2$.	2. Given.
3. $\angle 2 \cong \angle 1$.	3. If two sides of a triangle are congruent, the angles opposite those sides are congruent.
4. $\angle 3 \cong \angle 1$.	4. Transitive law.
5. $\angle EDA$ and $\angle 1$ are vertical angles.	5. Nonadjacent angles formed by intersecting lines segments are vertical angles.
6. $\angle EDA \cong \angle 1$.	6. Vertical angles are congruent.
7. $\angle EDA \cong \angle 3$.	7. Substitution.
8. $\overline{AE} \cong \overline{DE}$.	8. If two angles of a triangle are congruent, then sides opposite those angles are congruent.

Exercise 3.1 (page 89)

20. Given: Lines l_1 and l_2 with $\overleftrightarrow{AB}$ as a transversal. $\angle 1 \cong \angle 2$.

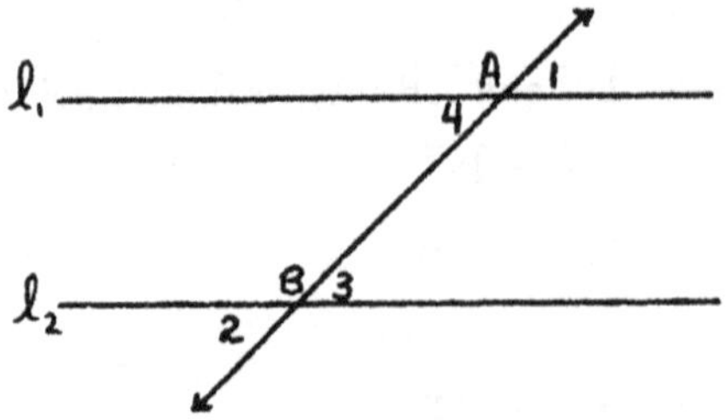

Prove: $l_1 \parallel l_2$.

STATEMENTS	REASONS
1. Lines l_1 and l_2 with $\overleftrightarrow{AB}$ as a transversal.	1. Given.
2. $\angle 1 \cong \angle 2$.	2. Given.
3. $\angle 2$ and $\angle 3$ are vertical angles.	3. Nonadjacent angles formed by intersecting lines segments are vertical angles.
4. $\angle 2 \cong \angle 3$.	4. Vertical angles are congruent.
5. $\angle 1$ and $\angle 4$ are vertical angles.	5. Nonadjacent angles formed by intersecting lines segments are vertical angles.
6. $\angle 1 \cong \angle 4$.	6. Vertical angles are congruent.
7. $\angle 4 \cong \angle 3$.	7. Substitution.
8. $l_1 \parallel l_2$.	8. If two lines are cut by a transversal so that a pair of alternate interior angles are congruent, then the lines are parallel.

22. Given: $\angle 2$ is supplementary to $\angle 3$.

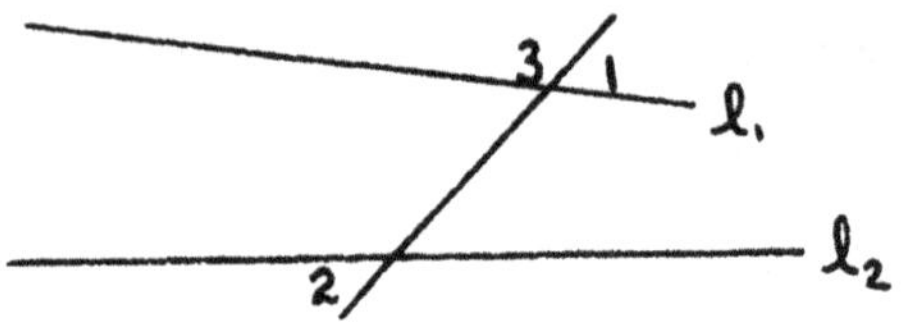

Prove: $l_1 \parallel l_2$.

STATEMENTS	REASONS
1. $\angle 2$ is supplementary to $\angle 3$.	1. Given.
2. $\angle 3$ and $\angle 1$ are supplementary.	2. Adjacent angles whose noncommon sides form a line are supplementary.
3. $\angle 2 \cong \angle 1$.	3. Supplements of the same angle are congruent.
4. $l_1 \parallel l_2$.	4. From exercise 20, if two lines are cut by a transversal so that alternate exterior angles are congruent, the lines are parallel.

24. Given: $\triangle EDC$ is isosceles with base $\overline{ED}$.
$\angle A \cong \angle 1$.

Prove: $\overline{AB} \parallel \overline{ED}$.

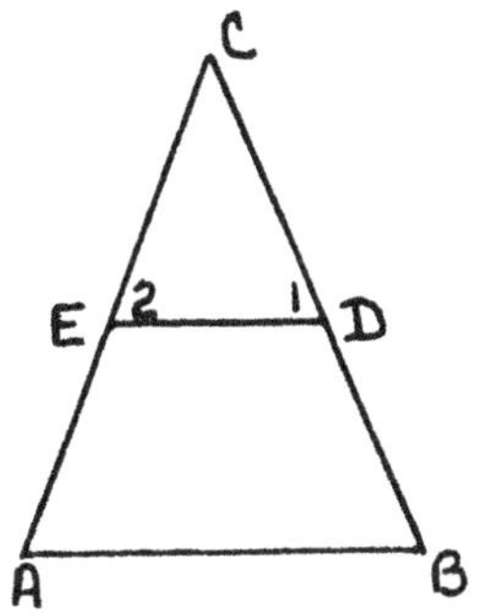

STATEMENTS	REASONS
1. $\triangle EDC$ is isosceles with base $\overline{ED}$.	1. Given.
2. $\angle A \cong \angle 1$.	2. Given.
3. $\overline{EC} \cong \overline{DC}$.	3. An isosceles triangle has two congruent sides.
4. $\angle 1 \cong \angle 2$.	4. If two sides of a triangle are congruent, then the angles opposite those sides are congruent.
5. $\angle A \cong \angle 2$.	5. Transitive law.
6. $\overline{AB} \parallel \overline{ED}$.	6. If two lines are cut by a transversal so that corresponding angles are congruent, the lines are parallel.

Exercise 3.2 (page 95)

26. Given: $\overline{AB} \parallel \overline{DC}$.

$\overline{AB} \cong \overline{CD}$.

Prove: $\angle A \cong \angle C$.

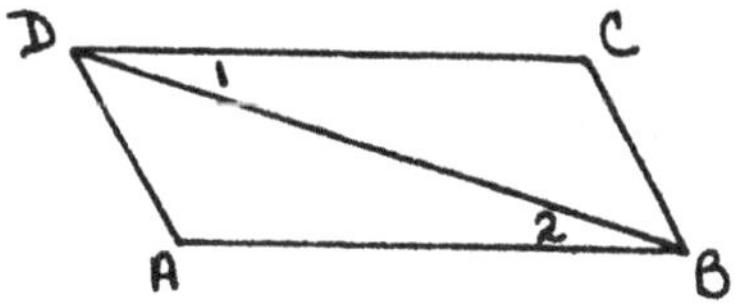

STATEMENTS	REASONS
1. $\overline{AB} \parallel \overline{DC}$.	1. Given.
2. $\overline{AB} \cong \overline{CD}$.	2. Given.
3. $\angle 2 \cong \angle 1$.	3. If two parallel lines are cut by a transversal, then all pairs of alternate interior angles are congruent.
4. $\overline{BD} \cong \overline{DB}$.	4. Reflexive law.
5. $\triangle ABD \cong \triangle CDB$.	5. SAS $\cong$ SAS.
6. $\angle A \cong \angle C$.	6. cpctc.

28. Given: $\overleftrightarrow{AB} \parallel \overleftrightarrow{CD}$.

$\overline{BC} \cong \overline{DC}$.

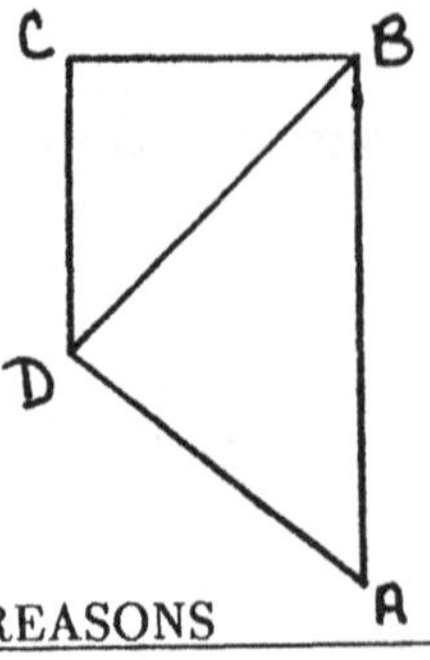

Prove: $\overleftrightarrow{BD}$ bisects $\angle CBA$.

STATEMENTS	REASONS
1. $\overleftrightarrow{AB} \parallel \overleftrightarrow{CD}$.	1. Given.
2. $\overline{BC} \cong \overline{DC}$.	2. Given.
3. $\angle 2 \cong \angle 3$.	3. If two sides of a triangle are congruent, then the angles opposite those sides are congruent.
4. $\angle 1 \cong \angle 2$.	4. If two parallel lines are cut by a transversal, all pairs of alternate interior angles are congruent.
5. $\angle 1 \cong \angle 3$.	5. Substitution.
6. $\overleftrightarrow{BD}$ bisects $\angle CBA$.	6. $\overleftrightarrow{BD}$ divides $\angle CBA$ into two congruent angles.

30. Given: $\triangle ABE$.

$\overline{AB} \parallel \overline{CD}$.

$\overline{CE} \cong \overline{DE}$.

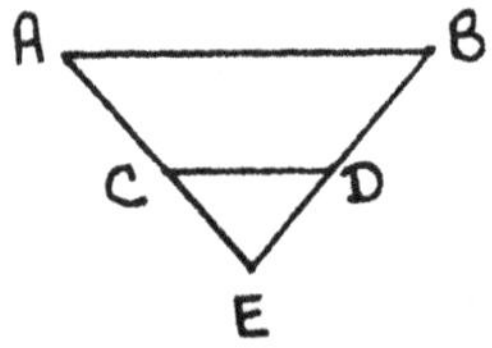

Prove: $\overline{AE} \cong \overline{BE}$.

STATEMENTS	REASONS
1. $\triangle ABE$.	1. Given.
2. $\overline{AB} \parallel \overline{CD}$.	2. Given.
3. $\overline{CE} \cong \overline{DE}$.	3. Given.
4. $\angle DCE \cong \angle CDE$.	4. If two sides of a triangle are congruent, then the angles opposite those sides are congruent.
5. $\angle A \cong \angle DCE$; $\angle B \cong \angle CDE$.	5. If two parallel lines are cut by a transversal, all pairs of corresponding angles are congruent.
6. $\angle A \cong \angle B$.	6. Substitution.
7. $\overline{AE} \cong \overline{BE}$.	7. If two angles of a triangle are congruent, then the sides opposite those angles are congruent.

32. Given: $l_1 \parallel l_2$.

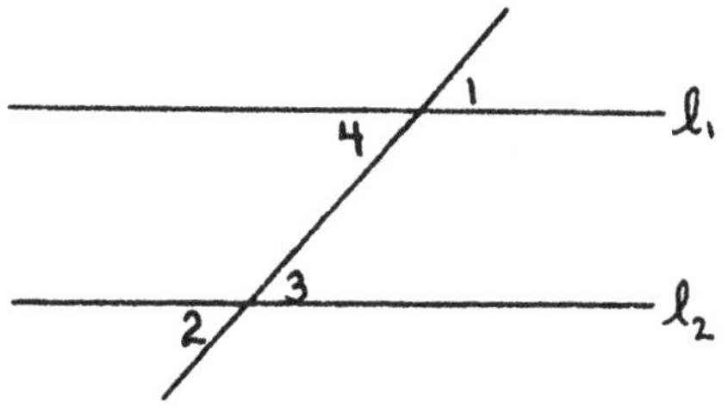

Prove: $\angle 1 \cong \angle 2$.

STATEMENTS	REASONS
1. $l_1 \parallel l_2$.	1. Given.
2. $\angle 3 \cong \angle 4$.	2. If two parallel lines are cut by a transversal, then all pairs of alternate interior angles are congruent.
3. $\angle 1$ and $\angle 4$ are vertical angles; $\angle 3$ and $\angle 2$ are vertical angles.	3. Nonadjacent angles formed by intersecting line segments are vertical angles.
4. $\angle 1 \cong \angle 4$; $\angle 3 \cong \angle 2$.	4. Vertical angles are congruent.
5. $\angle 1 \cong \angle 2$.	5. Substitution.

The proof for the other pair of alternate exterior angles is similar.

Exercise 3.3 (page 103)

20. Given: Lines $\overleftrightarrow{AC}$ and $\overleftrightarrow{ED}$ intersect at point B.
$\overline{AE} \perp \overline{AC}$; $\overline{CD} \perp \overline{AC}$.
$\overline{BE} \cong \overline{BD}$.

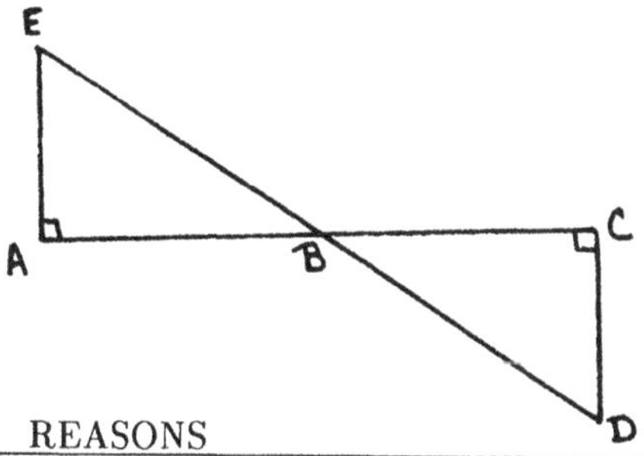

Prove: $\overline{AB} \cong \overline{CB}$.

STATEMENTS	REASONS
1. Lines $\overleftrightarrow{AC}$ and $\overleftrightarrow{ED}$ intersect at point B.	1. Given.
2. $\overline{AE} \perp \overline{AC}$; $\overline{CD} \perp \overline{AC}$.	2. Given.
3. $\overline{BE} \cong \overline{BD}$.	3. Given.
4. $\angle EAB$ is a right angle. $\angle DCB$ is a right angle.	4. Perpendicular lines meet at a right angle.
5. $\triangle ABE$ and $\triangle CBD$ are right triangles.	5. If a triangle contains a right angle, then it is a right triangle.
6. $\angle ABE$ and $\angle CBD$ are vertical angles.	6. Nonadjacent angles formed by intersecting lines are vertical angles.
7. $\angle ABE \cong \angle CBD$.	7. Vertical angles are congruent.
8. $\triangle ABE \cong \triangle CBD$.	8. HA $\cong$ HA.
9. $\overline{AB} \cong \overline{CB}$.	9. cpctc.

22. The proof is indirect.

Given: $\triangle ABC$.

Prove: $\triangle ABC$ can have at most one right angle.

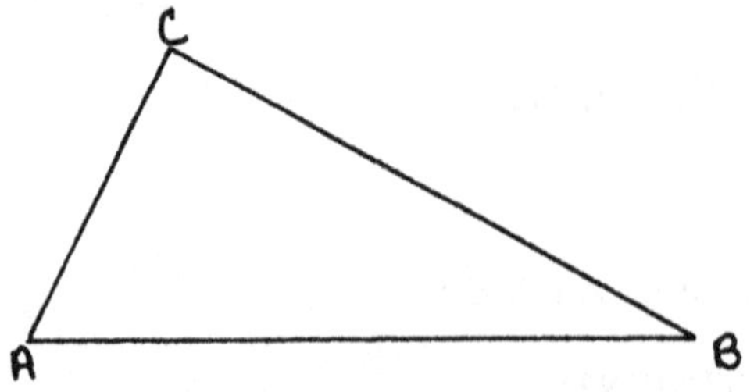

STATEMENTS	REASONS
1. Assume $\triangle ABC$ has more than one right angle.	1. Possible assumption.
2. The sum of the measures of the angles of $\triangle ABC$ is greater than 180°.	2. The sum of two right angles is 180°. The third angle will increase this sum to greater than 180°.
3. The sum of the measures of the angles of $\triangle ABC$ equals 180°.	3. Theorem 3.8.
4. The assumption that $\triangle ABC$ has more than one right angle is false.	4. It leads to a contradiction.
5. $\triangle ABC$ has at most one right angle.	5. It is the only remaining possibility.

24. Given: $\triangle ABC$ is a right triangle with $m(\angle C) = 90°$.

Prove: $\angle A$ and $\angle B$ are complementary.

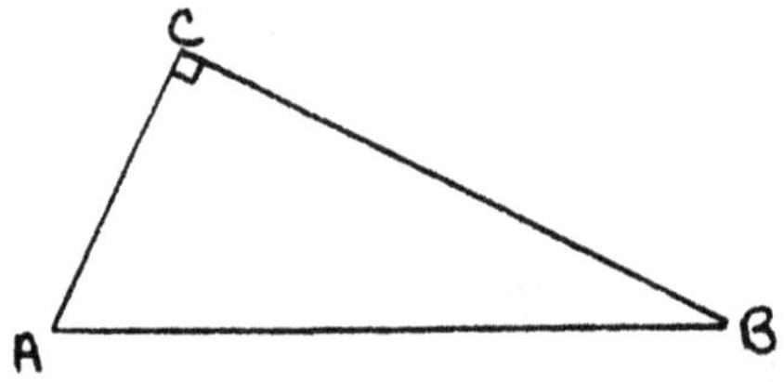

STATEMENTS	REASONS
1. $\triangle ABC$ is a right triangle with $m(\angle C) = 90°$.	1. Given.
2. $m(\angle A) + m(\angle B) + m(\angle C) = 180°$.	2. Theorem 3.8.
3. $m(\angle A) + m(\angle B) + 90° = 180°$.	3. Substitution.
4. $m(\angle A) + m(\angle B) = 90°$.	4. Equal quantities subtracted from equal quantities are equal.
5. $\angle A$ and $\angle B$ are complementary.	5. If the sum of the measures of two angles is 90°, the angles are complementary.

26. Given: $\triangle ABC$ and $\triangle DEF$ are right triangles with $\angle A$ and $\angle D$ are right angles.
$\overline{BC} \cong \overline{EF}$.
$\angle B \cong \angle E$.

Prove: $\triangle ABC \cong \triangle DEF$.

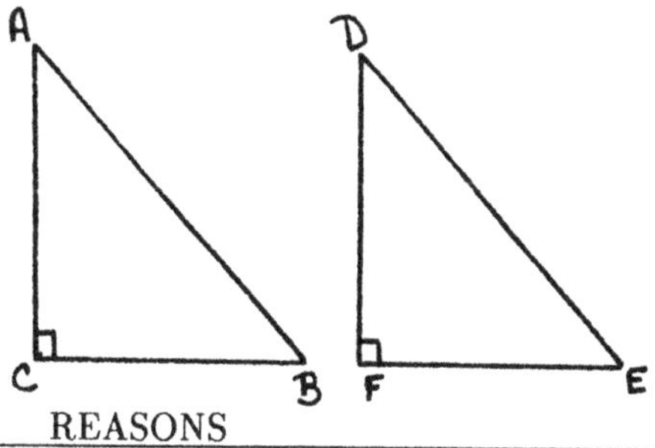

STATEMENTS	REASONS
1. $\triangle ABC$ and $\triangle DEF$ are right triangles with $\angle A$ and $\angle D$ are right angles.	1. Given.
2. $\overline{BC} \cong \overline{EF}$.	2. Given.
3. $\angle B \cong \angle E$.	3. Given.
4. $\angle A \cong \angle D$.	4. Right angles are congruent.
5. $\triangle ABC \cong \triangle DEF$.	5. Corollary 3.12.

28. Given: $m(\angle A) + m(\angle B) = m(\angle C)$.

Prove: $\triangle ABC$ is a right triangle.

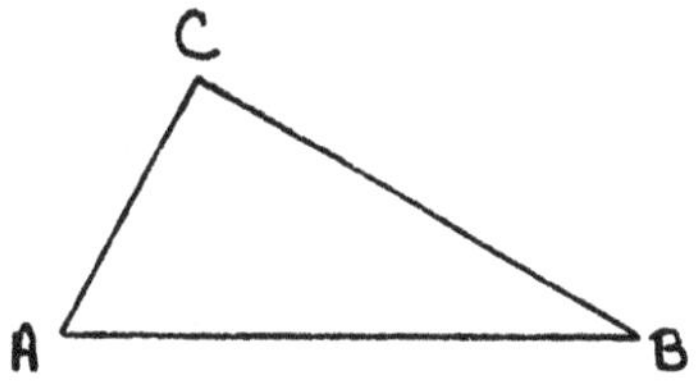

STATEMENTS	REASONS
1. $m(\angle A) + m(\angle B) = m(\angle C)$.	1. Given.
2. $m(\angle A) + m(\angle B) + m(\angle C) = m(\angle C) + m(\angle C)$.	2. Equal quantities added to equal quantities are equal.
3. $m(\angle A) + m(\angle B) + m(\angle C) = 180^\circ$.	3. Theorem 3.8.
4. $180^\circ = 2\,m(\angle C)$.	4. Substitution.
5. $90^\circ = m(\angle C)$.	5. Equal quantities divided by equal quantities are equal.
6. $\angle C$ is a right angle.	6. If an angle has measure of 90°, the angle is a right angle.
7. $\triangle ABC$ is a right triangle.	7. If a triangle has a right angle, it is a right triangle.

30. Given: $\triangle ABC$.

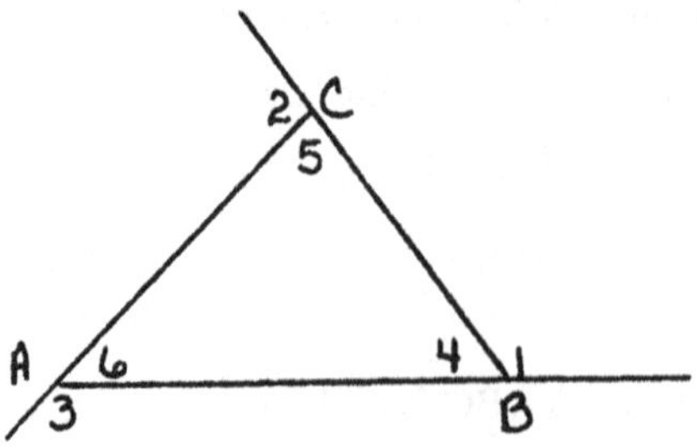

Prove: $m(\angle 1) + m(\angle 2) + m(\angle 3) + m(\angle 4) + m(\angle 5) + m(\angle 6) = 540^\circ$.

STATEMENTS	REASONS
1. $\triangle ABC$.	1. Given.
2. $m(\angle 1) + m(\angle 4) = 180^\circ$; $m(\angle 2) + m(\angle 5) = 180^\circ$; $m(\angle 3) + m(\angle 6) = 180^\circ$;	2. The measure of a straight angle is 180°.
3. $m(\angle 1) + m(\angle 2) + m(\angle 3) + m(\angle 4) + m(\angle 5) + m(\angle 6) = 540^\circ$.	3. Equal quantities added to equal quantities are equal.

Exercise 3.4 (page 109)

32. Show that the formula $a = 180^\circ - \frac{360^\circ}{n}$ is equivalent to the formula in Theorem 3.17.

$$a = 180^\circ - \frac{360^\circ}{n}$$

$$a = \frac{180^\circ}{1} \cdot \frac{n}{n} - \frac{360^\circ}{n}$$

$$a = \frac{180^\circ n}{n} - \frac{360^\circ}{n}$$

$$a = \frac{180^\circ n - 360^\circ}{n}$$

$$a = \frac{(n-2)180^\circ}{n}$$

36. Let n represent the number of sides of the polygon. Then, $(n-2)180^\circ$ represents the sum of the measures of the interior angles. By theorem 3.16, the sum of the measures of the exterior angles of a polygon is 360°. Thus,

$$(n-2)180^\circ = 360^\circ$$

$$180^\circ n - 360^\circ = 360^\circ$$

$$180^\circ n = 720^\circ$$

$$n = 4$$

Because $n = 4$, the polygon is a $4-$sided polygon.

Exercise 3.5 (page 113)

2. Given: $\angle A \ncong \angle B$.

Prove: $\overline{BC} \ncong \overline{AC}$.

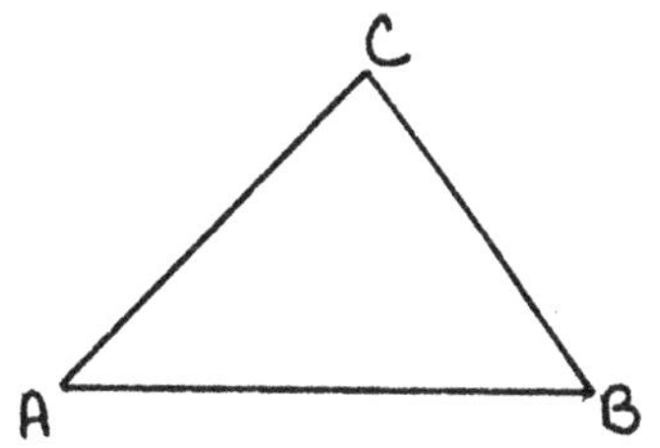

STATEMENTS	REASONS
1. Assume that $\overline{BC} \cong \overline{AC}$.	1. Possible assumption.
2. $\angle A \cong \angle B$.	2. If two sides of a triangle are congruent, then the angles opposite those sides are congruent.
3. $\angle A \ncong \angle B$.	3. Given.
4. The assumption that $\overline{BC} \cong \overline{AC}$ is false.	4. It leads to a contradiction.
5. $\overline{BC} \ncong \overline{AC}$.	5. It is the only remaining possibility.

4. Given: Lines l_1 and l_2 with $\overleftrightarrow{AB}$ as a transversal.
$\angle 1$ and $\angle 2$ are not supplementary.

Prove: $l_1 \nparallel l_2$.

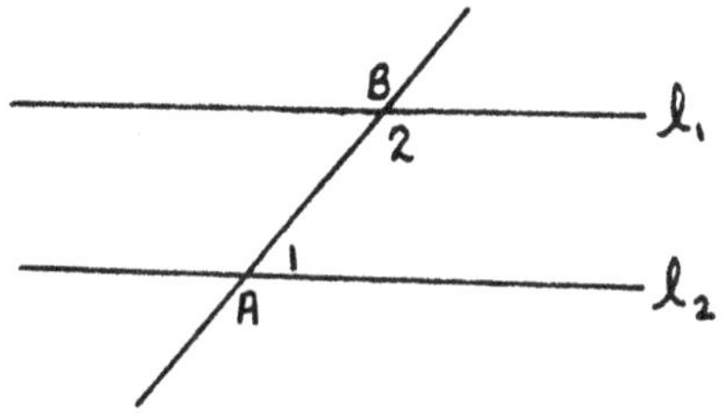

STATEMENTS	REASONS
1. Assume $l_1 \parallel l_2$.	1. Possible assumption.
2. $\angle 1$ and $\angle 2$ are supplementary.	2. If two parallel lines are cut by a transversal, then interior angles on the same side of the transversal are supplementary.
3. $\angle 1$ and $\angle 2$ are not supplementary.	3. Given.
4. The assumption that $l_1 \parallel l_2$ is false.	4. It leads to a contradiction.
5. $l_1 \nparallel l_2$.	5. It is the only remaining possibility.

6. Given: Lines l_1 and l_2 with $\overleftrightarrow{AB}$ as a transversal.

$l_1 \not\parallel l_2$.

Prove: $\angle 1 \ncong \angle 2$.

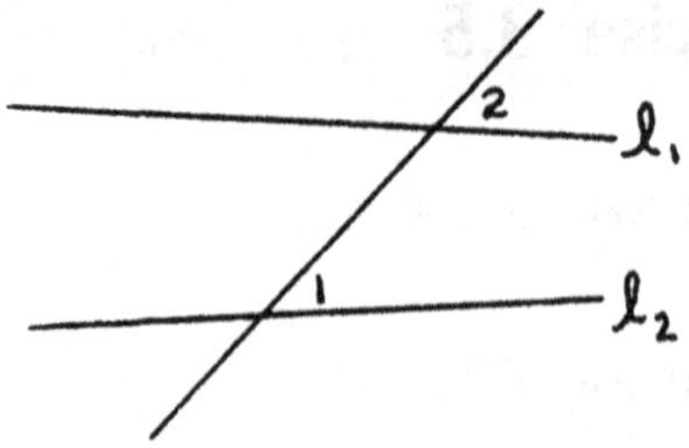

STATEMENTS	REASONS
1. Assume $\angle 1 \cong \angle 2$.	1. Possible assumption.
2. $l_1 \parallel l_2$.	2. If two lines are cut by a transversal, so that a pair of corresponding angles are congruent, the lines are parallel.
3. $l_1 \not\parallel l_2$.	3. Given.
4. The assumption that $\angle 1 \cong \angle 2$ is false.	4. It leads to a contradiction.
5. $\angle 1 \ncong \angle 2$.	5. It is the only remaining possibility.

8. Given: The sum of the measures of the angles of the polygon ABCDE is not 180°.

Prove: Polygon ABCDE is not a triangle.

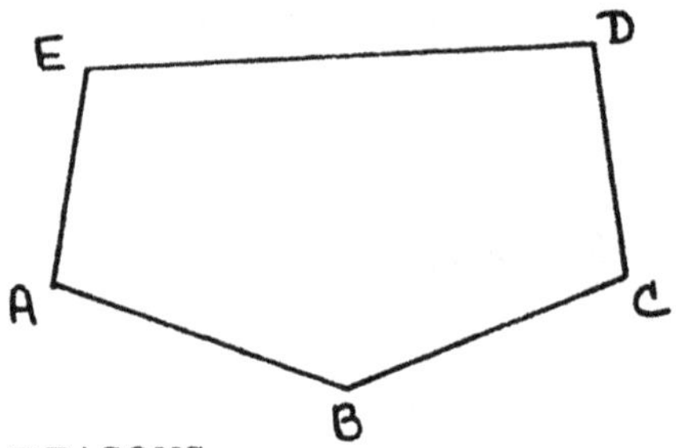

STATEMENTS	REASONS
1. Polygon ABCDE is a triangle.	1. Possible assumption.
2. The sum of the measures of the angles of the triangle ABCDE is 180°.	2. Theorem 3.8.
3. The sum of the measures of the angles of the polygon ABCDE is not 180°.	3. Given.
4. The assumption that polygon ABCDE is a triangle is false.	4. It leads to a contradiction.
5. The polygon ABCDE is not a triangle.	5. It is the only remaining possibility.

10. Given: $l_1 \parallel l_3$.

$l_2 \parallel l_3$.

Prove: $l_1 \parallel l_2$.

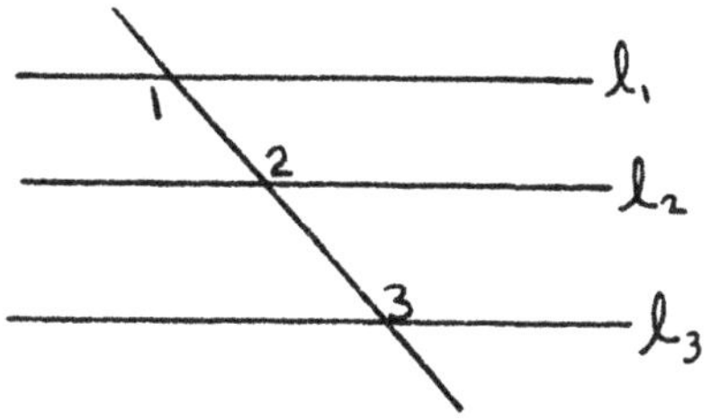

STATEMENTS	REASONS
1. Assume $l_1 \nparallel l_2$.	1. Possible assumption.
2. $\angle 1 \ncong \angle 2$.	2. Theorem 3.21.
3. $l_1 \parallel l_3$; $l_2 \parallel l_3$.	3. Given.
4. $\angle 1 \cong \angle 3$.	4. If parallel lines are cut by a transversal, then all pairs of alternate interior angles are congruent.
5. $\angle 2 \cong \angle 3$.	5. If parallel lines are cut by a transversal, then all pairs of corresponding angles are congruent.
6. $\angle 1 \cong \angle 2$.	6. Substitution.
7. The assumption that $l_1 \nparallel l_2$ is false.	7. It leads to a contradiction in steps 2 and 6.
8. $l_1 \parallel l_2$.	8. It is the only remaining possibility.

Chapter Three Review Exercises (page 114)

16. Given: $\overline{AC}$ and $\overline{ED}$ bisect each other.

Prove: $\overline{AE} \parallel \overline{CD}$.

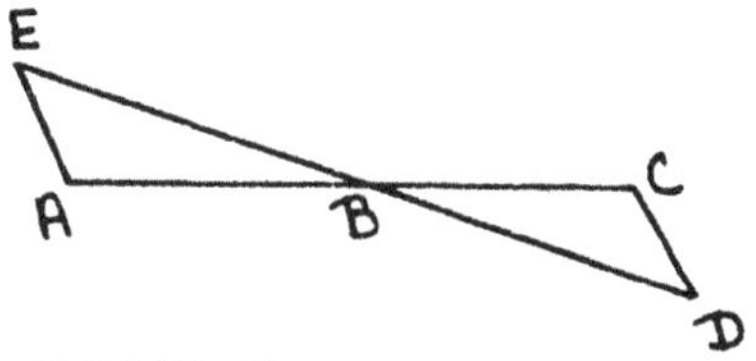

STATEMENTS	REASONS
1. $\overline{AC}$ and $\overline{ED}$ bisect each other.	1. Given.
2. $\overline{AB} \cong \overline{CB}$; $\overline{EB} \cong \overline{DB}$.	2. Bisectors divide a line segment into two congruent parts.
3. $\angle EBA$ and $\angle DBC$ are vertical angles.	3. Nonadjacent angles formed by intersecting lines are vertical angles.
4. $\angle EBA \cong \angle DBC$.	4. Vertical angles are congruent.
5. $\triangle ABE \cong \triangle CBD$.	5. SAS $\cong$ SAS

Continued on the next page.

6. $\angle E \cong \angle D$.	6. cpctc.
7. $\overline{AE} \parallel \overline{CD}$.	7. If two lines are cut by a transversal so that alternate interior angles are congruent, the lines are parallel.

30. Given: The sum of the measures of the angles of the polygon ABCDE is not 360°.

Prove: Polygon ABCDE is not a quadrilateral.

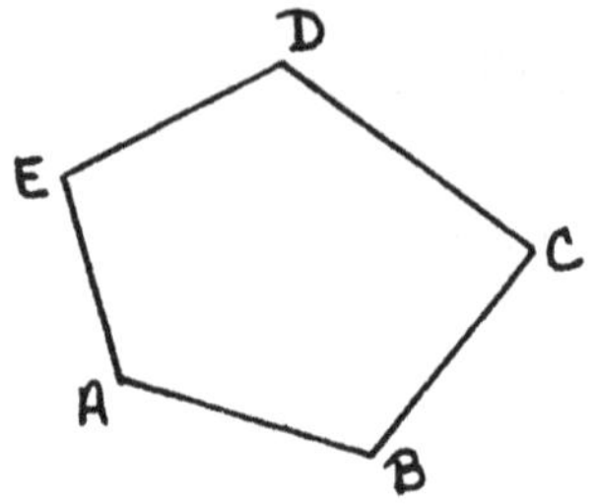

STATEMENTS	REASONS
1. Assume the polygon ABCDE is a quadrilateral.	1. Possible assumption.
2. The sum of the measures of the angles of the polygon ABCDE is 360°.	2. A quadrilateral has four sides, so the sum of the angles is $(4-2)180^\circ = (2)180^\circ = 360^\circ$.
3. The sum of the measures of the angles of the polygon ABCDE is not 360°.	3. Given.
4. The assumption is false.	4. It leads to a contradiction.
5. Polygon ABCDE is not a quadrilateral.	5. It is the only remaining possibility.

Exercise 4.1 (page 123)

18. Given: ▱ABCD.

Prove: $\angle D \cong \angle B$.

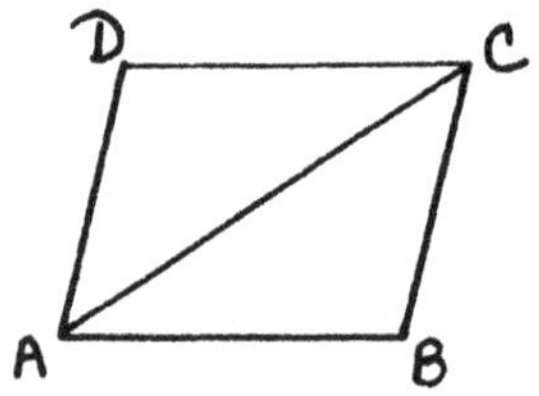

STATEMENTS	REASONS
1. ABCD is a parallelogram.	1. Given.
2. $\triangle ACD \cong \triangle CAB$.	2. A diagonal of a parallelogram divides the parallelogram into two congruent triangles.
3. $\angle D \cong \angle B$.	3. cpctc.

20. Given: $\overline{AB} \parallel \overline{CD}$.

$\overline{AB} \cong \overline{CD}$.

Prove: ABCD is a parallelogram.

Construction: Diagonal $\overline{AC}$.

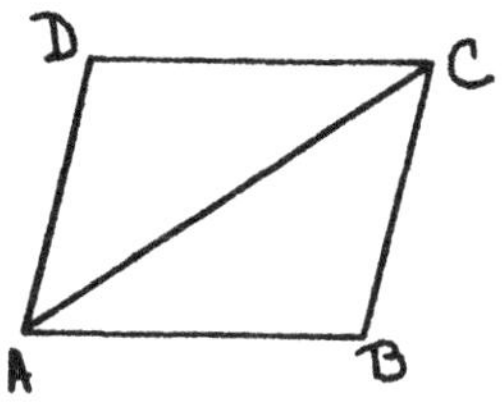

STATEMENTS	REASONS
1. $\overline{AB} \parallel \overline{CD}$.	1. Given.
2. $\overline{AB} \cong \overline{CD}$.	2. Given.
3. $\angle ACD \cong \angle CAB$.	3. If two parallel lines are cut by a transversal, then alternate interior angles are congruent.
4. $\overline{CA} \cong \overline{AC}$.	4. Reflexive law.
5. $\triangle ACD \cong \triangle CAB$.	5. SAS $\cong$ SAS.
6. $\overline{AD} \cong \overline{CB}$.	6. cpctc.
7. ABCD is a parallelogram.	7. If both pairs of opposite sides of a quadrilateral are congruent, then the quadrilateral is a parallelogram.

24. Given: ABCD is a parallelogram in which $\overline{LN}$ bisects diagonal $\overline{AC}$ at point M.

Prove: $\overline{AC}$ bisects $\overline{LN}$.

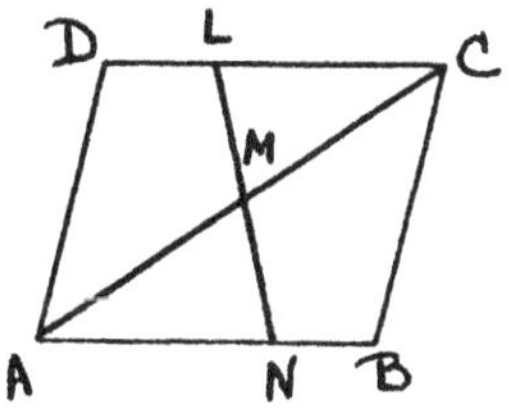

STATEMENTS	REASONS
1. ABCD is a parallelogram in which $\overline{LN}$ bisects diagonal $\overline{AC}$ at point M.	1. Given.
2. $\overline{AM} \cong \overline{CM}$.	2. A bisector divides a line segment into two congruent segments.
3. $\angle AMN \cong \angle CML$.	3. Vertical angles are congruent.
4. $\overline{AB} \parallel \overline{DC}$.	4. Opposite sides of a parallelogram are parallel.
5. $\angle BAC \cong \angle DCA$.	5. If two parallel lines are cut by a transversal, alternate interior angles are congruent.

Continued on the next page.

STATEMENTS	REASONS
6. $\triangle AMN \cong \triangle CML$.	6. ASA ≅ ASA.
7. $\overline{LM} \cong \overline{NM}$.	7. cpctc.
8. $\overline{AC}$ bisects $\overline{LN}$.	8. If a line divides a given segment into two congruent segments, it bisects the given segment.

26. Given: Quadrilateral ABCD, with its diagonals intersecting at point M.

$\angle 1 \cong \angle 2$.

$\overline{AB} \cong \overline{CD}$.

Prove: $\overline{MD} \cong \overline{MB}$.

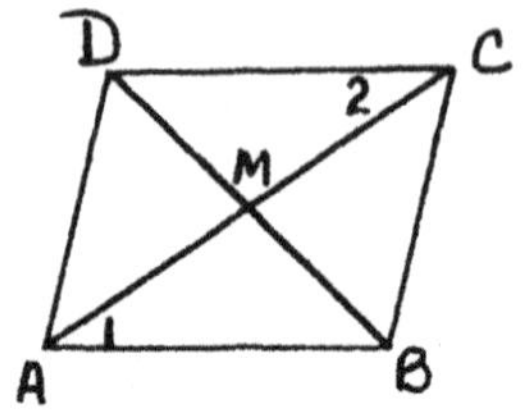

STATEMENTS	REASONS
1. Quadrilateral ABCD, with its diagonals intersecting at point M.	1. Given.
2. $\angle 1 \cong \angle 2$.	2. Given.
3. $\overline{AB} \cong \overline{CD}$.	3. Given.
4. $\overline{AB} \parallel \overline{CD}$.	4. If two lines are cut by a transversal so that alternate interior angles are congruent, the lines are parallel.
5. ABCD is a parallelogram.	5. If opposite sides of a quadrilateral are both parallel and congruent, the quadrilateral is a parallelogram.
6. $\overline{MD} \cong \overline{MB}$.	6. The diagonals of a parallelogram bisect each other.

28. Given: ▱ABCD.

$\overline{AM} \perp \overline{DB}$.

$\overline{CN} \perp \overline{DB}$.

Prove: AMCN is a ▱.

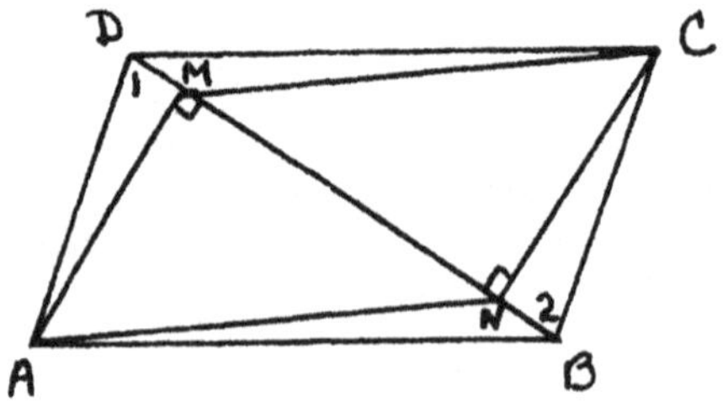

The proof is on the next page.

STATEMENTS	REASONS
1. ▱ABCD.	1. Given.
2. $\overline{AD} \parallel \overline{BC}$.	2. Opposite sides of a parallelogram are parallel.
3. $\angle 1 \cong \angle 2$.	3. If two parallel lines are cut by a transversal, alternate interior angles are equal.
4. $\overline{AD} \cong \overline{BC}$.	4. Opposite sides of a parallelogram are congruent.
5. $\overline{AM} \perp \overline{DB}$; $\overline{CN} \perp \overline{DB}$.	5. Given.
6. $\angle$DMA and $\angle$BNC are right angles.	6. Perpendicular lines meet and form right angles.
7. $\triangle$AMD and $\triangle$BNC are right triangles.	7. If a triangle has a right angle, it is a right triangle.
8. $\triangle AMD \cong \triangle CNB$.	8. HA $\cong$ HA.
9. $\overline{AM} \cong \overline{CN}$.	9. cpctc.
10. $\overline{AM} \parallel \overline{CN}$.	10. If two lines are each perpendicular to a third line, they are parallel to each other.
11. AMCN is a ▱.	11. If opposite sides of a quadrilateral are both parallel and congruent, the quadrilateral is a parallelogram.

Exercise 4.2 (page 131)

14. Given: ABCD is a rectangle with $m(\angle A) = 90°$.

Prove: $\angle$B, $\angle$C, and $\angle$D are right angles.

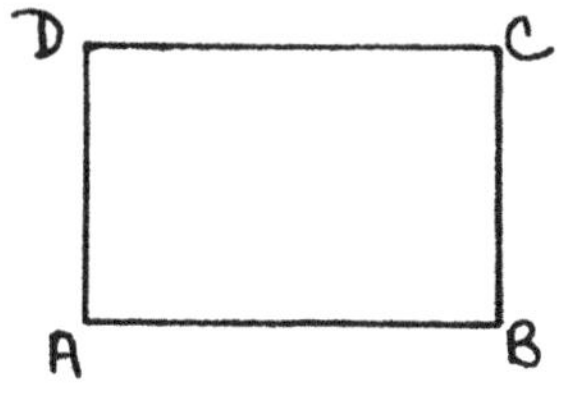

STATEMENTS	REASONS
1. ABCD is a rectangle with $m(\angle A) = 90°$.	1. Given.
2. $m(\angle A) = m(\angle C) = 90°$.	2. Because a rectangle is a parallelogram, its nonconsecutive angles are congruent.
3. $\angle B \cong \angle D$.	3. Same as reason 2.

Continued on the next page.

4. $\angle A$ is supplementary to $\angle D$.	4. Consecutive angles of a parallelogram are supplementary.
5. $m(\angle A) + m(\angle D) = 180°$.	5. The sum of the measures of supplementary angles is 180°.
6. $90° + m(\angle D) = 180°$.	6. Substitution.
7. $m(\angle D) = 90°$.	7. If equal quantities are subtracted from equal quantities, the results are equal.
8. $m(\angle B) = 90°$.	8. Substitution.
9. $\angle B$, $\angle C$, and $\angle D$ are right angles.	9. If an angle has measure of 90°, the angle is a right angle.

16. Given: ABCD is a rhombus with $\overline{AC}$ and $\overline{BD}$ diagonals.

Prove: $\angle 1 \cong \angle 2$.

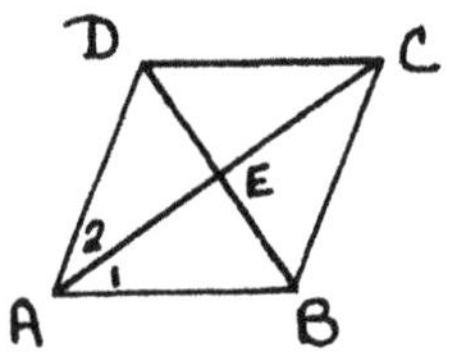

STATEMENTS	REASONS
1. ABCD is a rhombus with $\overline{AC}$ and $\overline{BD}$ diagonals.	1. Given.
2. $\overline{AD} \cong \overline{AB}$.	2. All sides of a rhombus are congruent.
3. $\overline{AC}$ bisects $\overline{BD}$.	3. The diagonals of a rhombus bisect each other.
4. $\overline{DE} \cong \overline{BE}$.	4. If a segment is bisected, it is divided into two congruent parts.
5. $\overline{AE} \cong \overline{AE}$.	5. Reflexive law.
6. $\triangle ADE \cong \triangle ABE$.	6. SSS $\cong$ SSS.
7. $\angle 1 \cong \angle 2$.	7. cpctc.

18. Given: ABCD is a quadrilateral with E, F, G, and H midpoints of their respective sides.

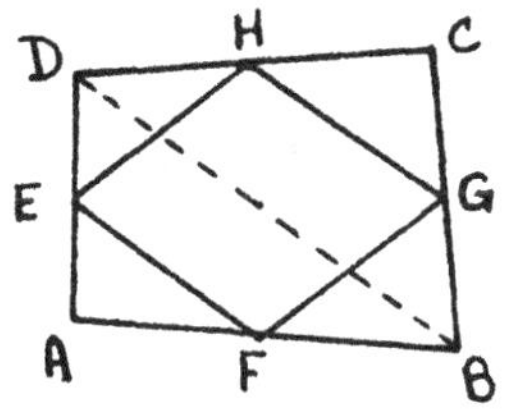

Prove: EFGH is a ▱.

Construction: Diagonal $\overline{DB}$.

STATEMENTS	REASONS
1. ABCD is a quadrilateral with E, F, G, and H as midpoints of their respective sides.	1. Given.
2. $\overline{EF} \parallel \overline{DB}$; $m(\overline{EF}) = \frac{1}{2}m(\overline{DB})$. $\overline{GH} \parallel \overline{DB}$; $m(\overline{GH}) = \frac{1}{2}m(\overline{DB})$.	2. Theorem 4.17.
3. $\overline{EF} \parallel \overline{GH}$.	3. If two lines are each parallel to a third line, they are parallel to each other.
4. $m(\overline{EF}) = m(\overline{GH})$.	4. Substitution.
5. $\overline{EF} \cong \overline{GH}$.	5. If line segments have the same measure, they are congruent.
6. EFGH is a ▱.	6. If a pair of opposite sides of a quadrilateral are parallel and congruent, then the quadrilateral is a parallelogram.

20. Given: $\triangle$ABC is an equilateral triangle. D, E, and F midpoints of their respective sides.

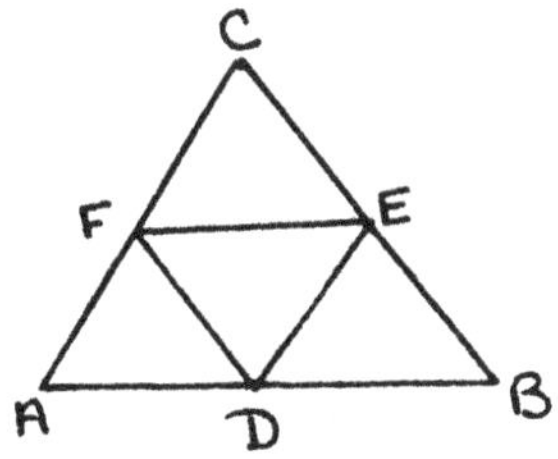

Prove: $\triangle$DEF is equilateral.

STATEMENTS	REASONS
1. $\triangle$ABC is an equilateral triangle.	1. Given.
2. D, E, and F as midpoints of their respective sides.	2. Given.
3. $\overline{AB} \cong \overline{BC} \cong \overline{AC}$.	3. All sides of an equilateral triangle are congruent.

Continued on the next page.

4. $m(\overline{AB}) = m(\overline{BC}) = m(\overline{AC})$.

4. Congruent segments have equal measuure.

5. $\frac{1}{2}m(\overline{AB}) = \frac{1}{2}m(\overline{BC}) = \frac{1}{2}m(\overline{AC})$.

5. Equal quantities multiplied by equal quantities are equal.

6. $m(\overline{DE}) = \frac{1}{2}m(\overline{AC})$; $m(\overline{EF}) = \frac{1}{2}m(\overline{AB})$; $m(\overline{DF}) = \frac{1}{2}m(\overline{BC})$.

6. Theorem 4.17.

7. $m(\overline{EF}) = m(\overline{DF}) = m(\overline{DE})$.

7. Substitution.

8. $\overline{EF} \cong \overline{DF} \cong \overline{DE}$.

8. If line segments have equal measure, they are congruent.

9. $\triangle$ DEF is an equilateral triangle.

9. If a triangle has three congruent sides, the triangle is equialteral.

22. Given: ABCD is a rhombus with E, F, G, and H as midpoints of their respective sides.

Prove: EFGH is a rectangle.

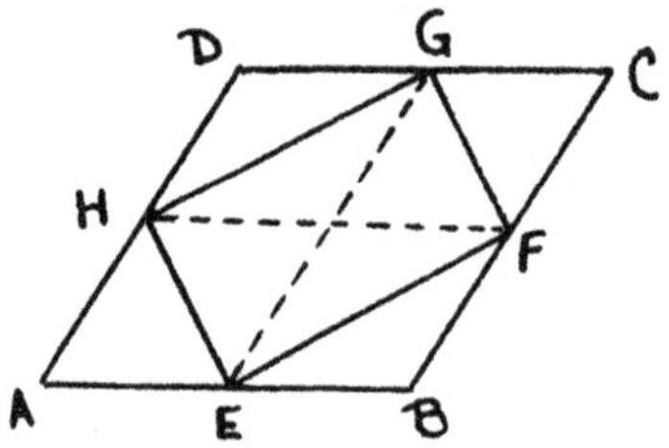

STATEMENTS	REASONS
1. ABCD is a rhombus with E, F, G, and H as midpoints of their respective sides.	1. Given.
2. EFGH is a $\square$.	2. Theorem 4.18.
3. $\overline{AD} \cong \overline{BC}$; $\overline{AB} \cong \overline{DC}$.	3. Opposite sides of a rhombus are congruent.
4. $m(\overline{AD}) = m(\overline{BC})$; $m(\overline{AB}) = m(\overline{DC})$.	4. Congruent segments have equal measuure.
5. $\frac{1}{2}m(\overline{AD}) = \frac{1}{2}m(\overline{BC})$; $\frac{1}{2}m(\overline{AB}) = \frac{1}{2}m(\overline{DC})$.	5. Equal quantities multiplied by equal quantities are equal.
6. $m(\overline{HD}) = \frac{1}{2}m(\overline{AD})$; $m(\overline{FC}) = \frac{1}{2}m(\overline{BC})$; $m(\overline{AE}) = \frac{1}{2}m(\overline{AB})$; $m(\overline{DG}) = \frac{1}{2}m(\overline{DC})$.	6. Theorem 1.5.
7. $m(\overline{HD}) = m(\overline{FC})$; $m(\overline{AE}) = m(\overline{DG})$.	7. Substitution.
8. $\overline{HD} \cong \overline{FC}$; $\overline{AE} \cong \overline{DG}$.	8. If line segments have equal measure, they are congruent.
9. $\overline{HD} \parallel \overline{FC}$; $\overline{AE} \parallel \overline{DG}$.	9. Opposite sides of a rhombus are parallel.

Continued on the next page.

STATEMENTS	REASONS
10. HFCD is a ▱; EGDA is a ▱.	10. If a pair of opposite sides are parallel and congruent, the quadrilateral is a parallelogram.
11. $\overline{DC} \cong \overline{HF}$; $\overline{AD} \cong \overline{EG}$.	11. Opposite sides of a parallelogram are congruent.
12. $\overline{DC} \cong \overline{AD}$.	12. Adjacent sides of a rhombus are congruent.
13. $\overline{HF} \cong \overline{EG}$.	13. Substitution.
14. EFGH is a rectangle.	14. If the diagonals of a parallelogram are congruent, the parallelogram is a rectangle.

24. Given: $\triangle$ ABC is a right triangle with $m(\angle B) = 90°$.

L and M are midpoints of the two legs.

N is the midpoint of the hypotenuse.

Prove: LBMN is a rectangle.

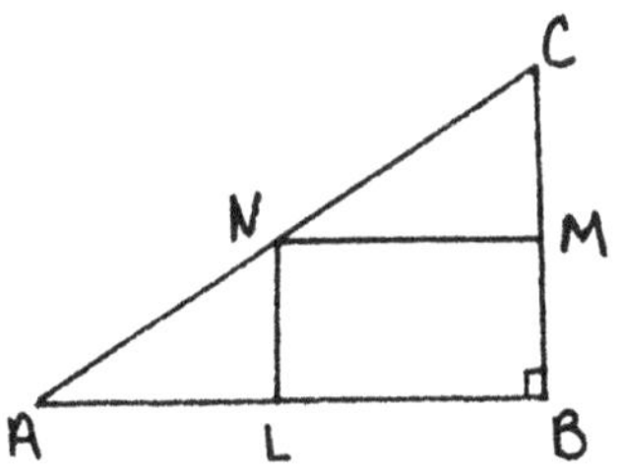

STATEMENTS	REASONS
1. $\triangle$ ABC is a right triangle with $m(\angle B) = 90°$.	1. Given.
2. L and M are midpoints of the two legs.	2. Given.
3. N is the midpoint of the hypotenuse.	3. Given.
4. $\overline{MN} \parallel \overline{BL}$; $\overline{LN} \parallel \overline{BM}$.	4. If a line joins the midpoints of two sides of a triangle, then the line is parallel to the third side.
5. LBMN is a ▱.	5. If both pairs of opposite sides are parallel, then the quadrilateral is a parallelogram.
6. $\angle$B is a right angle.	6. An angle with measure of 90° is a right angle.
7. LBMN is a rectangle.	7. A rectangle is a parallelogram with one right angle.

Exercise 4.3 (page 138)

14. Given: ABCD is an isosceles trapezoid with $\overline{AD} \cong \overline{BC}$ and $\overline{DC} \parallel \overline{AB}$.

E and F are midpoints of their respective sides.

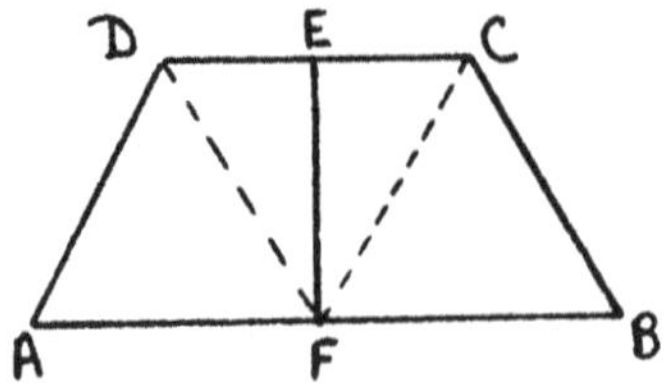

Prove: $\overline{EF} \perp \overline{DC}$; $\overline{EF} \perp \overline{AB}$.

Construction: Line segment $\overline{DF}$ and $\overline{CF}$.

STATEMENTS	REASONS
1. ABCD is an isosceles trapezoid with $\overline{AD} \cong \overline{BC}$ and $\overline{DC} \parallel \overline{AB}$.	1. Given.
2. E is the midpoint of $\overline{DC}$ and F is the midpoint of $\overline{AB}$.	2. Given.
3. $\angle A \cong \angle B$.	3. Base angles of an isosceles trapezoid are congruent.
4. $\overline{AF} \cong \overline{BF}$.	4. A midpoint divides a line segment into two congruent parts.
5. $\triangle ADF \cong \triangle BCF$.	5. SAS $\cong$ SAS.
6. $\overline{DF} \cong \overline{CF}$.	6. cpctc.
7. $\triangle DFC$ is isosceles.	7. If a triangle has two congruent sides, it is isosceles.
8. $\overline{FE} \cong \overline{FE}$.	8. Reflexive law.
9. $\overline{DE} \cong \overline{CE}$.	9. A midpoint divides a line segment into congruent segments.
10. $\triangle FED \cong \triangle FEC$.	10. SSS $\cong$ SSS.
11. $\angle DFE \cong \angle CFE$.	11. cpctc.
12. $\overline{FE}$ bisects $\angle DFC$.	12. Definition of an angle bisector.
13. $\overline{EF} \perp \overline{DC}$.	13. The bisector of the vertex angle of an isosceles triangle is the perpendicular bisector of the base of the triangle.
14. $\overline{EF} \perp \overline{AB}$.	14. If a line is perpendicualr to one of two parallel lines, it is perpendicular to the other line also.

16. Given: Quadrilateral ABCD is a ▱.

Points E, G, H, and I are midpoints of $\overline{DE}$, $\overline{AE}$; $\overline{BE}$; $\overline{CE}$ respectively.

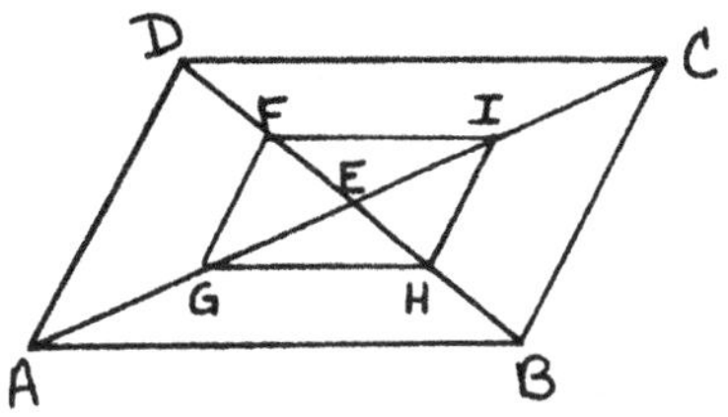

Prove: Quadrilateral FGHI is a ▱.

STATEMENTS	REASONS
1. ABCD is a ▱ with F, G, H, and I as midpoints of $\overline{DE}$; $\overline{AE}$; $\overline{BE}$; $\overline{CE}$ respectively.	1. Given.
2. $\overline{AD} \cong \overline{BC}$; $\overline{AD} \parallel \overline{BC}$.	2. Opposite sides of a parallelogram are congruent and parallel.
3. $\frac{1}{2}m(\overline{AD}) = \frac{1}{2}m(\overline{BC})$.	3. Equal quantities multiplied by equal quantities are equal.
4. $\overline{GF} \parallel \overline{AD}$; $m(\overline{GF}) = \frac{1}{2}m(\overline{AD})$. $\overline{HI} \parallel \overline{BC}$; $m(\overline{HI}) = \frac{1}{2}m(\overline{BC})$.	4. If a line segment joins the midpoints of two sides of a triangle, then that line is parallel to and congruent to one-half the measure of the third side.
5. $\overline{AD} \parallel \overline{HI}$; $\overline{GF} \parallel \overline{HI}$.	5. If two distinct lines are parallel to a third line, then they are parallel to each other.
6. $\overline{GF} \cong \overline{HI}$.	6. Substitution.
7. FGHI is a ▱.	7. If a pair of opposite sides are parallel and congruent, the quadrilateral is a parallelogram.

Chapter Four Review Exercises (page 139)

10. Given: $\square$ABCD with E, F, G, and H midpoints of their respective sides.

Prove: $\overline{HG} \cong \overline{FE}$.

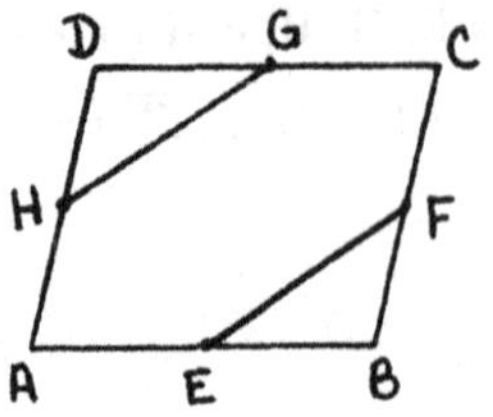

STATEMENTS	REASONS
1. ABCD is a $\square$ with E, F, G, and H as midpoints of their respective sides.	1. Given.
2. $\angle D \cong \angle B$.	2. Nonconsecutive angles of a parallelogram are congruent.
3. $\overline{DC} \cong \overline{AB}$; $\overline{AD} \cong \overline{BC}$.	3. Opposites sides of a parallelogram are congruent.
4. $\overline{DG} \cong \overline{BE}$; $\overline{DH} \cong \overline{BF}$.	4. The line segment bisector theorem.
5. $\triangle HDG \cong \triangle FBE$.	5. SAS $\cong$ SAS.
6. $\overline{HG} \cong \overline{FE}$.	6. cpctc.

Exercise 5.1 (page 154)

38. Given: $\square$ABCD.
Point E is between B and C.

Prove: Area($\triangle$ AED) $= \frac{1}{2}$Area($\square$ABCD).

Construction: Draw $\overline{EF} \perp \overline{AD}$.

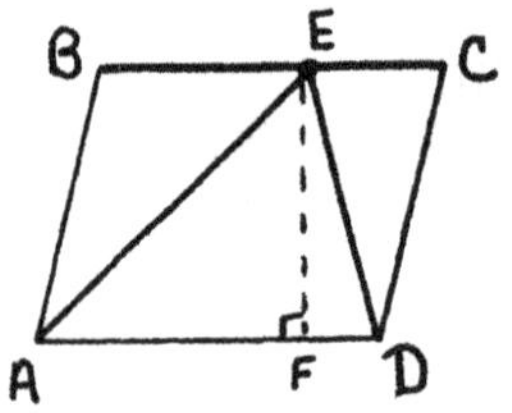

STATEMENTS	REASONS
1. ABCD is a $\square$ with point E between B and C.	1. Given.
2. Area($\square$ABCD) $=$ m($\overline{AD}$) $\cdot$ m($\overline{FE}$).	2. Theorem 5.1.
3. Area($\triangle$ ADE) $= \frac{1}{2}$m($\overline{AD}$) $\cdot$ m($\overline{FE}$).	3. Theorem 5.2.
4. Area($\triangle$ ADE) $= \frac{1}{2}$Area($\square$ABCD).	4. Substitution.

42. Given: Isosceles right $\triangle ABC$, with $\overline{AB}$ as the hypotenuse and $\angle C$ as a right angle.

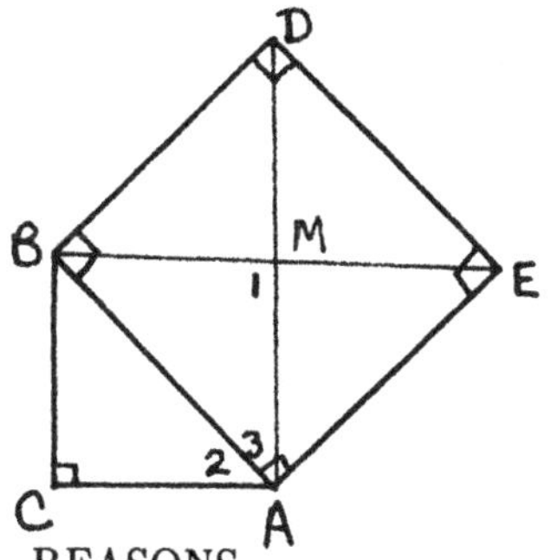

Prove: Area($\triangle ABC$) $= \frac{1}{4}$Area($\square$ ABDE).

STATEMENTS	REASONS
1. Isosceles right $\triangle ABC$, with $\overline{AB}$ as the hypotenuse and $\angle C$ as a right angle.	1. Given.
2. ABDE is a square with diagonals $\overline{BE}$ and $\overline{AD}$.	2. By construction.
3. $\overline{BE} \cong \overline{AD}$.	3. The diagonals of a square are congruent.
4. $m(\overline{BE}) = m(\overline{AD})$.	4. Congruent segments have equal measuure.
5. $\frac{1}{2}m(\overline{BE}) = \frac{1}{2}m(\overline{AD})$.	5. Equal quantities multiplied by equal quantities are equal.
6. $\overline{BE}$ and $\overline{AD}$ bisect each other.	6. Diagonals of a square bisect each other.
7. $\overline{BM} \cong \overline{ME}$; $\overline{AM} \cong \overline{MD}$.	7. A bisector divides the line segment into two congruent segments.
8. $\frac{1}{2}m(\overline{BE}) = m(\overline{BM})$; $\frac{1}{2}m(\overline{AD}) = m(\overline{AM})$.	8. Theorem 1.5.
9. $m(\overline{BM}) = m(\overline{AM})$.	9. Substitution.
10. $\overline{BM} \cong \overline{AM}$.	10. If the measures of two line segments are equal, the line segments are congruent.
11. $\overline{BM} \cong \overline{ME} \cong \overline{AM} \cong \overline{MD}$.	11. Line segments that are congruent to congruent segments are congruent.
12. $\overline{AB} \cong \overline{BD} \cong \overline{DE} \cong \overline{AE}$.	12. All sides of a square are congruent.
13. $\triangle ABM \cong \triangle BMD \cong \triangle DME \cong \triangle EMA$.	13. SSS $\cong$ SSS.
14. Area($\triangle ABM$) $= \frac{1}{4}$Area($\square$ ABDE).	14. Postulate 5.4.
15. $\angle 1$ is a right angle.	15. Diagonals of a square meet and form right angles.
16. $\overline{AB} \cong \overline{AB}$.	16. Reflexive law.
17. $m(\angle 2) = 45°$.	17. Base angles of an isosceles right triangle have measure of 45°.
18. $m(\angle 3) = 45°$.	18. The diagonals of a square bisect the 90° angles of the square.
19. $\triangle ABC \cong \triangle ABM$.	19. HA $\cong$ HA.
20. Area($\triangle ABC$) $= \frac{1}{4}$Area($\square$ ABDE).	20. Substitution.

44. Given: ▱ABCD with diagonals $\overline{AC}$ and $\overline{DB}$.

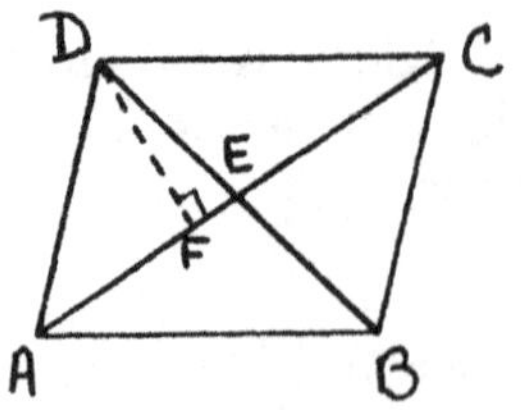

Prove: Area(△ AEB) = Area(△ CEB) = Area(△ CED) = Area(△ AED).

Construction: $\overline{DF} \perp \overline{AB}$.

STATEMENTS	REASONS
1. ▱ABCD with diagonals $\overline{AC}$ and $\overline{DB}$.	1. Given.
2. $\overline{AB} \cong \overline{CD}$; $\overline{AB} \parallel \overline{DC}$. $\overline{AD} \cong \overline{CB}$; $\overline{AD} \parallel \overline{CB}$.	2. Opposite sides of a ▱ are congruent and parallel.
3. $\overline{DE} \cong \overline{BE}$; $\overline{AE} \cong \overline{CE}$.	3. Diagonals of a ▱ bisect each other.
4. △ CED ≅ △ AEB.	4. SSS ≅ SSS.
5. Area(△ CED) = Area(△ AEB).	5. Postulate 5.2.
6. △ AED ≅ △ CEB.	6. SSS ≅ SSS.
7. Area(△ AED) = Area(△ CEB).	7. Postulate 5.2.
8. $m(\overline{AE}) = m(\overline{CE})$.	8. Congruent line segments have equal measure.
9. $\overline{DF} \perp \overline{AC}$.	9. By construction.
10. $\overline{DF}$ is an altitude of △ ADE and △ CED.	10. Definition of an altitude.
11. $m(\overline{DF})$ is the height of △ ADE and △ CED.	11. Definition 5.1.
12. $m(\overline{DF}) = m(\overline{DF})$.	12. Reflexive law.
13. Area(△ AED) = Area(△ CED).	13. The triangles have bases of equal measure and equal heights so the areas are equal.
14. Area(△ AEB) = Area(△ CEB) = Area(△ CED) = Area(△ AED).	14. Substitution.

46. Given: $\triangle$ ABC is a right triangle with m($\angle$C) = 90°.

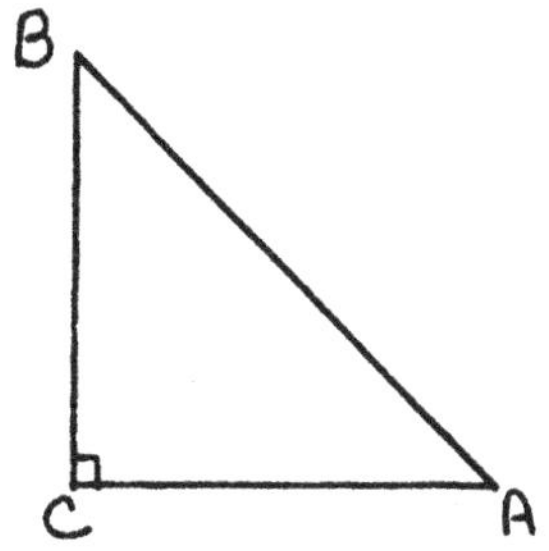

Prove: Area($\triangle$ ABC) = $\frac{1}{2}$m($\overline{AC}$) · m($\overline{BC}$).

STATEMENTS	REASONS
1. $\triangle$ ABC is a right triangle with m($\angle$C) = 90°.	1. Given.
2. Area of a triangle = $\frac{1}{2}$(base)(height).	2. Theorem 5.2.
3. Area($\triangle$ ABC) = $\frac{1}{2}$m($\overline{AC}$) · m($\overline{BC}$).	3. Substitution.

Exercise 5.3 (page 168)

38. Sketch of the proof: Given the isosceles triangle to the right, the area is

$A = \frac{1}{2}x \cdot x$ or $A = \frac{1}{2}x^2$.

Because the hypotenuse is $x\sqrt{2}$, the square of the hypotenuse is $2x^2$. Thus, the area of the triangle is $\frac{1}{4}$ the square of the hypotenuse.

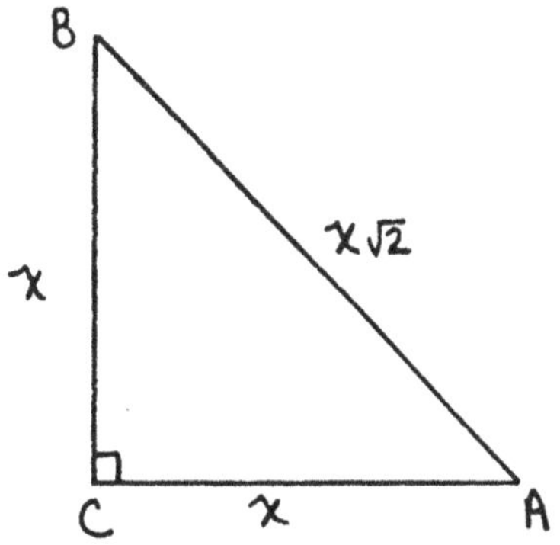

Exercise 5.4 (page 174)

34. Given: Isosceles right $\triangle$ ABC with m($\overline{AB}$) = m($\overline{BC}$) = x.

Prove: m($\overline{AC}$) = $x\sqrt{2}$.

Use the Pythagorean theorem:

$$x^2 + x^2 = [\mathrm{m}(\overline{AC})]^2$$

$$2x^2 = [\mathrm{m}(\overline{AC})]^2$$

$$\sqrt{2x^2} = \mathrm{m}(\overline{AC})$$

$$x\sqrt{2} = \mathrm{m}(\overline{AC})$$

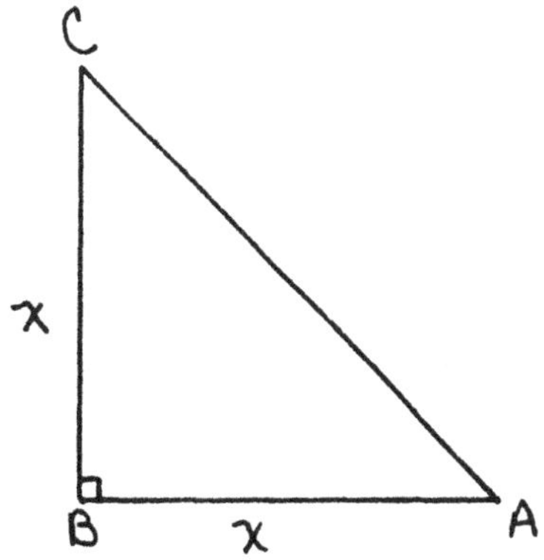

Exercise 6.1 (page 198)

62. Given: $\frac{a}{b} = \frac{c}{d}$.

Prove: $\frac{a-b}{b} = \frac{c-d}{d}$.

STATEMENTS	REASONS
1. $\frac{a}{b} = \frac{c}{d}$.	1. Given.
2. $\frac{a}{b} - 1 = \frac{c}{d} - 1$.	2. Equal quantities subtracted from equal quantities are equal.
3. $\frac{b}{b} = 1$; $\frac{d}{d} = 1$.	3. Any nonzero quantity divided by itself is 1.
4. $\frac{a}{b} - \frac{b}{b} = \frac{c}{d} - \frac{d}{d}$.	4. Substitution.
5. $\frac{a-b}{b} = \frac{c-d}{d}$.	5. Substitution.

Exercise 6.2 (page 206)

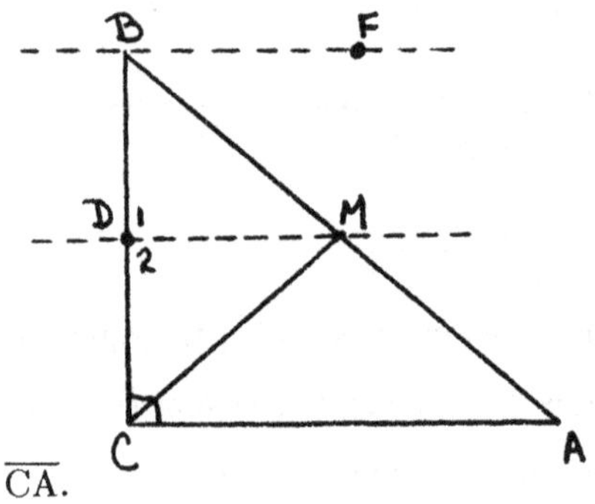

50. Given: Right $\triangle$ ABC with $\overline{DC} \perp \overline{AC}$.

M is the midpoint of $\overline{AB}$.

Prove: $\overline{AM} \cong \overline{MB} \cong \overline{MC}$.

Construction: Draw $\overline{DM}$ and $\overline{BF}$ parallel to $\overline{CA}$.

STATEMENTS	REASONS
1. Right $\triangle$ ABC ; M is the midpoint of $\overline{AB}$.	1. Given.
2. $\overline{AM} \cong \overline{MB}$.	2. A midpoint of a segment divides the segment into two congruent segments.
3. $\overline{DM} \parallel \overline{CA}$; $\overline{BF} \parallel \overline{CA}$.	3. By construction.
4. $\overline{BD} \cong \overline{DC}$.	4. Theorem 4.21.
5. $\overline{DM} \cong \overline{DM}$.	5. Reflexive law.
6. $\overline{DC} \perp \overline{AC}$.	6. Given.
7. $\overline{DC} \perp \overline{DM}$.	7. Theorem 3.4.
8. $\angle 1 \cong \angle 2$.	8. Definition 1.18.
9. $\triangle$ BDM $\cong$ $\triangle$ CDM.	9. SAS $\cong$ SAS.
10. $\overline{MB} \cong \overline{MC}$.	10. cpctc.
11. $\overline{AM} \cong \overline{MB} \cong \overline{MC}$.	11. Steps 2 and 10.

52. Given: Lines $\overleftrightarrow{AD}$ and $\overleftrightarrow{BE}$ intersect at point C.

$\overline{AB} \parallel \overline{DE}$.

Prove: $\frac{m(\overline{AB})}{m(\overline{DE})} = \frac{m(\overline{AC})}{m(\overline{DC})}$.

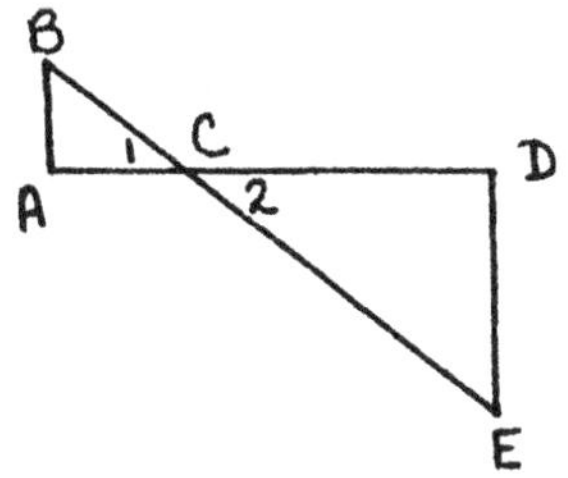

STATEMENTS	REASONS
1. Lines $\overleftrightarrow{AD}$ and $\overleftrightarrow{BE}$ intersect at point C.	1. Given.
2. $\overline{AB} \parallel \overline{DE}$.	2. Given.
3. $\angle B \cong \angle E$.	3. If two parallel lines are cut by a transversal, alternate interior angles are congruent.
4. $\angle 1$ and $\angle 2$ are vertical angles.	4. Nonadjacent angles formed by intersecting lines are vertical angles.
5. $\angle 1 \cong \angle 2$.	5. Vertical angles are congruent.
6. $\triangle ACB \sim \triangle DCE$.	6. If two angles of one triangle are congruent to two angles of another triangle, the triangles are similar.
7. $\frac{m(\overline{AB})}{m(\overline{DE})} = \frac{m(\overline{AC})}{m(\overline{DC})}$.	7. Corresponding sides of similar triangles are in proportion.

54. Given: $\triangle ABE$.

$\angle 1 \cong \angle A$.

Prove: $\frac{m(\overline{AE})}{m(\overline{AC})} = \frac{m(\overline{BE})}{m(\overline{BD})}$.

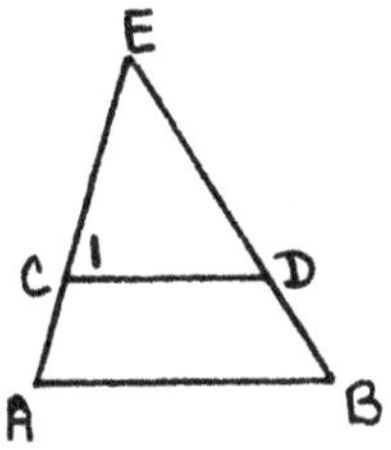

STATEMENTS	REASONS
1. $\triangle ABE$; $\angle 1 \cong \angle A$.	1. Given.
2. $\overline{CD} \parallel \overline{AB}$.	2. If two lines are cut by a transversal so that a pair of corresponding angles are congruent, the lines are parallel.
3. $\frac{m(\overline{CE})}{m(\overline{AC})} = \frac{m(\overline{ED})}{m(\overline{BD})}$.	3. Theorem 6.9.
4. $\frac{m(\overline{CE}) + m(\overline{AC})}{m(\overline{AC})} = \frac{m(\overline{ED}) + m(\overline{BD})}{m(\overline{BD})}$.	4. Theorem 6.5.
5. $m(\overline{CE}) + m(\overline{AC}) = m(\overline{AE})$. $m(\overline{ED}) + m(\overline{BD}) = m(\overline{BE})$.	5. Postulate 1.10.
6. $\frac{m(\overline{AE})}{m(\overline{AC})} = \frac{m(\overline{BE})}{m(\overline{BD})}$.	6. Substitution.

56. Given: $\triangle ABC$ and $\triangle ADE$ are right triangles.

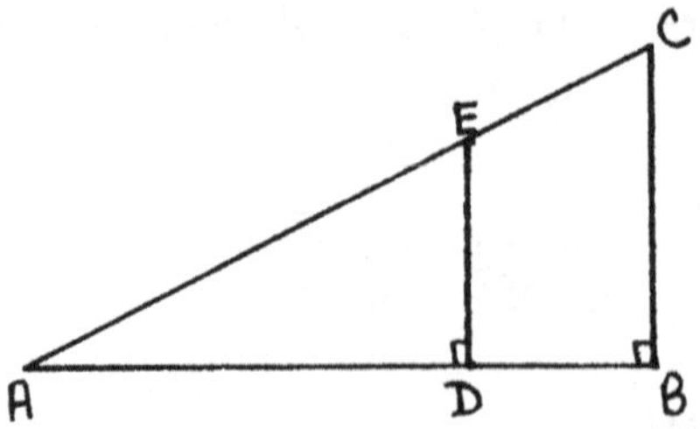

Prove: $[m(\overline{AD})][m(\overline{BC})] = [m(\overline{AB})][m(\overline{DE})]$.

STATEMENTS	REASONS
1. $\triangle ABC$ and $\triangle ADE$ are right triangles.	1. Given.
2. $m(\angle EDA) = m(\angle CBA) = 90^\circ$.	2. The measure of a right angle is 90°.
3. $\angle A \cong \angle A$.	3. Reflexive law.
4. $\triangle ADE \sim \triangle ABC$.	4. If two angles of one triangle are congruent to two angles of another triangle, the triangles are similar.
5. $\dfrac{m(\overline{AD})}{m(\overline{AB})} = \dfrac{m(\overline{DE})}{m(\overline{BC})}$.	5. Corresponding sides of similar triangles are in proportion.
6. $[m(\overline{AD})][m(\overline{BC})] = [m(\overline{AB})][m(\overline{DE})]$.	6. Theorem 6.1.

Exercise 6.3 (page 217)

32. Given: $\triangle ABC$ and $\triangle DEF$ are isosceles triangles with vertex angles C and F congruent.

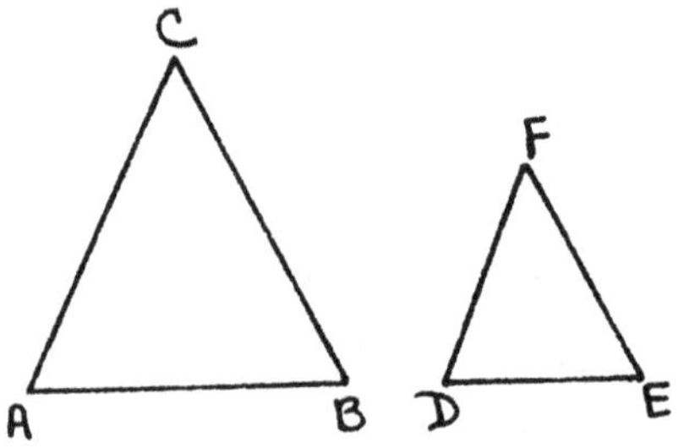

Prove: $\triangle ABC \sim \triangle DEF$.

STATEMENTS	REASONS
1. $\triangle ABC$ and $\triangle DEF$ are isosceles triangles with vertex angles C and F congruent.	1. Given.
2. $\angle A \cong \angle B$; $\angle D \cong \angle E$.	2. Base angles of an isosceles triangle are congruent.
3. $m(\angle A) = m(\angle B)$; $m(\angle D) = m(\angle E)$; $m(\angle C) = m(\angle F)$.	3. Congruent angles have equal measure.
4. $m(\angle A) + m(\angle B) + m(\angle C) = 180^\circ$; $m(\angle D) + m(\angle E) + m(\angle F) = 180^\circ$.	4. The sum of the measures of the angle of a triangle is 180°.

Continued on the next page.

5. $m(\angle A) + m(\angle B) + m(\angle C) = m(\angle D) + m(\angle E) + m(\angle F)$. — 5. Substitution.

6. $m(\angle A) + m(\angle B) = m(\angle D) + m(\angle E)$. — 6. Equal quantities subtracted from equal quantities are equal.

7. $m(\angle A) + m(\angle A) = m(\angle D) + m(\angle D)$. — 7. Substitution.

8. $2m(\angle A) = 2m(\angle D)$. — 8. Postulate 1.9.

9. $m(\angle A) = m(\angle D)$. — 9. Equal quantities divided by equal quantities are equal.

10. $\angle A \cong \angle D$. — 10. Angles with equal measure are congruent.

11. $\triangle ABC \sim \triangle DEF$. — 11. If two angles of one triangle are congruent to two angles of another triangle, the triangles are similar.

34. Given: $\triangle ABC \sim \triangle DEF$.

$\overline{CG}$ and $\overline{FH}$ are medians.

Prove: $\frac{m(\overline{CG})}{m(\overline{FH})} = \frac{m(\overline{AC})}{m(\overline{DF})}$.

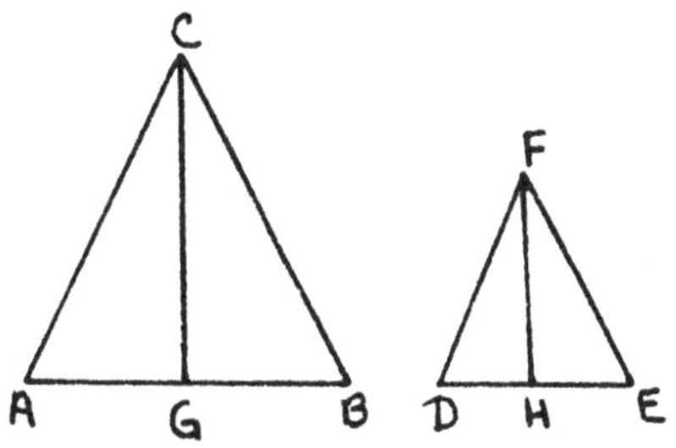

STATEMENTS	REASONS
1. $\triangle ABC \sim \triangle DEF$; $\overline{CG}$ and $\overline{FH}$ are medians.	1. Given.
2. $\angle A \cong \angle D$.	2. Corresponding angles in similar triangles are congruent.
3. $\frac{m(\overline{AC})}{m(\overline{DF})} = \frac{m(\overline{AB})}{m(\overline{DE})}$.	3. Corresponding sides of similar triangles are in proportion.
4. G is the midpoint of $\overline{AB}$. H is the midpoint of $\overline{DE}$.	4. A median of a triangle is a line drawn from a vertex to the midpoint of the opposite side.
5. $m(\overline{AG}) = \frac{(\overline{AB})}{2}$; $m(\overline{DH}) = \frac{m(\overline{DE})}{2}$.	5. Theorem 1.5.
6. $2\,m(\overline{AG}) = m(\overline{AB})$; $2\,m(\overline{DH}) = m(\overline{DE})$.	6. Equal quantities multiplied by equal quantities are equal.
7. $\frac{m(\overline{AC})}{m(\overline{DF})} = \frac{2\,m(\overline{AG})}{2\,m(\overline{DH})}$.	7. Substitution.

Continued on the next page.

8. $\dfrac{m(\overline{AC})}{m(\overline{DF})} = \dfrac{m(\overline{AG})}{m(\overline{DH})}$.	8. Postulate 1.9.
9. $\triangle AGC \sim \triangle DHF$.	9. Theorem 6.14.
10. $\dfrac{m(\overline{CG})}{m(\overline{FH})} = \dfrac{m(\overline{AC})}{m(\overline{DF})}$.	10. Corresponding sides of similar triangles are in proportion.

36. Given: $\triangle ABC \sim \triangle DEF$.

$\overline{CG}$ and $\overline{FH}$ are altitudes.

Prove: $\dfrac{m(\overline{CG})}{m(\overline{FH})} = \dfrac{m(\overline{AC})}{m(\overline{DF})}$.

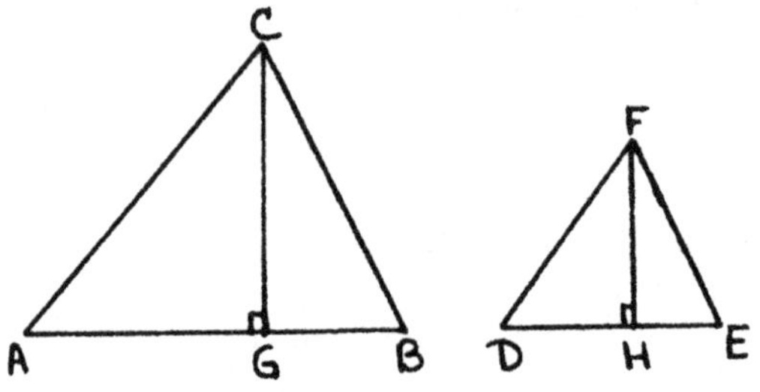

STATEMENTS	REASONS
1. $\triangle ABC \sim \triangle DEF$.	1. Given.
2. $\overline{CG}$ and $\overline{FH}$ are altitudes.	2. Given.
3. $\angle A \cong \angle D$.	3. Corresponding angles in similar triangles are congruent.
4. $\overline{CG} \perp \overline{AB}$; $\overline{FH} \perp \overline{DE}$.	4. Altitudes are segments drawn from a vertex, perpendicular to the opposite side.
5. $\angle CGA$ is a right angle; $\angle FHD$ is a right angle.	5. Perpendicular lines meet and form right angles.
6. $\angle CGA \cong \angle FHD$.	6. All right angles are congruent.
7. $\triangle AGC \sim \triangle DHF$.	7. If two angles of one triangle are congruent to two angles of a second triangle, the triangles are similar.
8. $\dfrac{m(\overline{CG})}{m(\overline{FH})} = \dfrac{m(\overline{AC})}{m(\overline{DF})}$.	8. Corresponding sides of similar triangles are in proportion.

38. Given: ABCD and EFGH are squares.

Prove: ABCD $\sim$ EFGH.

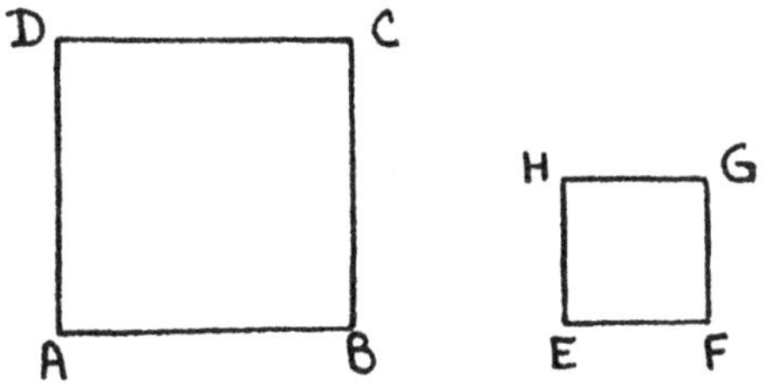

STATEMENTS	REASONS
1. ABCD and EFGH are squares.	1. Given.
2. $m(\angle A) = m(\angle B) = m(\angle C) = m(\angle D) = m(\angle E) = m(\angle F) = m(\angle G) = m(\angle H) = 90°$.	2. Every angle in a square has measure of 90°.
3. $\overline{AB} \cong \overline{BC} \cong \overline{CD} \cong \overline{DA}$; $\overline{EF} \cong \overline{FG} \cong \overline{GH} \cong \overline{HE}$.	3. All sides of a square are congruent.
4. $\dfrac{m(\overline{AB})}{m(\overline{EF})} = \dfrac{m(\overline{BC})}{m(\overline{FG})} = \dfrac{m(\overline{CD})}{m(\overline{GH})} = \dfrac{m(\overline{DA})}{m(\overline{HE})}$.	4. Equal quantities divided by equal quantities are equal.
5. ABCD $\sim$ EFGH.	5. Definition 6.6.

40. Given: $\triangle ABC \sim \triangle GHI$.

$\triangle DEF \sim \triangle GHI$.

Prove: $\triangle ABC \sim \triangle DEF$.

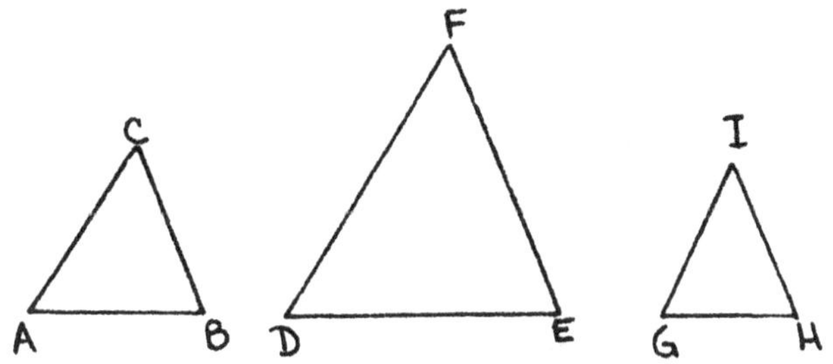

STATEMENTS	REASONS
1. $\triangle ABC \sim \triangle GHI$.	1. Given.
2. $\triangle DEF \sim \triangle GHI$.	2. Given.
3. $\angle A \cong \angle G$; $\angle B \cong \angle H$; $\angle D \cong \angle G$: $\angle E \cong \angle H$.	3. Corresponding angles in similar triangles are congruent.
4. $\angle A \cong \angle D$; $\angle B \cong \angle E$.	4. Substitution.
5. $\triangle ABC \sim \triangle DHF$.	5. If two angles of one triangle are congruent to two angles of a second triangle, the triangles are similar.

Exercise 7.1 (page 246)

30. PART I

Given: Circle O and circle O′ are congruent.
$\overline{AB}$ is a diameter of circle O and
$\overline{CD}$ is a diameter of circle O′.

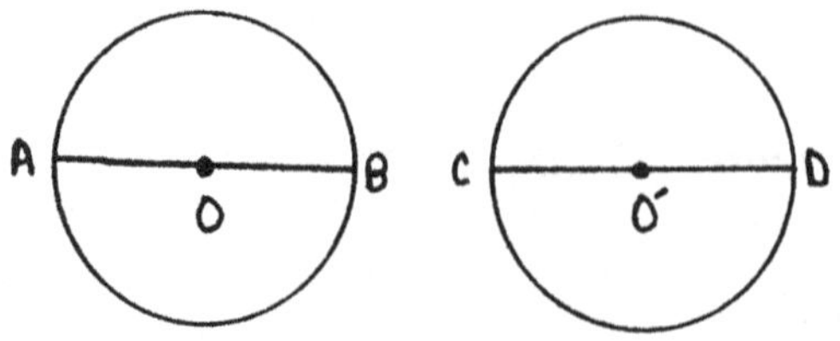

Prove: $\overline{AB} \cong \overline{CD}$.

STATEMENTS	REASONS
1. Circle O and circle O′ are congruent.	1. Given.
2. $\overline{AB}$ is a diameter of circle O and $\overline{CD}$ is a diameter of circle O′.	2. Given.
3. $\overline{OB} \cong \overline{O'D}$.	3. If two circles are congruent, their radii are congruent.
4. $m(\overline{OB}) = m\left(\overline{O'D}\right)$.	4. Congruent segments have equal measure.
5. $2\,m(\overline{OB}) = 2\,m\left(\overline{O'D}\right)$.	5. Equal quantities muliplied by equal quantities are equal.
6. $2\,m(\overline{OB}) = m(\overline{AB})$. $2\,m\left(\overline{O'D}\right) = m(\overline{CD})$.	6. Theorem 7.1.
7. $m(\overline{AB}) = m(\overline{CD})$.	7. Substitution.
8. $\overline{AB} \cong \overline{CD}$.	8. Segments with equal measure are congruent.

PART II

Given: $\overline{AB} \cong \overline{CD}$.

Prove: Circle O and circle O′ are congruent.

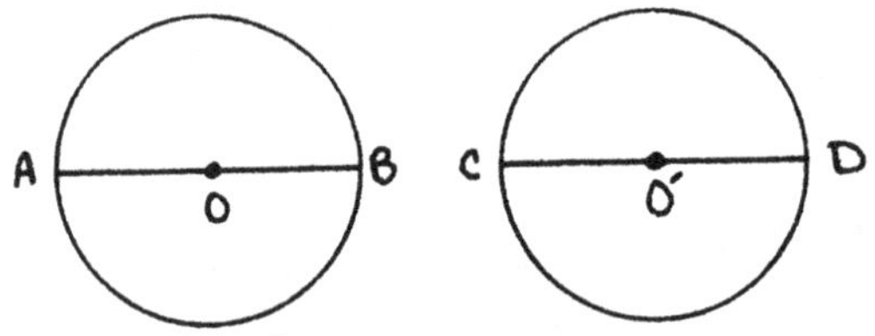

STATEMENTS	REASONS
1. $\overline{AB} \cong \overline{CD}$.	1. Given.
2. $m(\overline{AB}) = m(\overline{CD})$.	2. Congruent segments have equal measure.
3. $2\,m(\overline{OB}) = m(\overline{AB})$. $2\,m\left(\overline{O'D}\right) = m(\overline{CD})$.	3. Theorem 7.1.
4. $2\,m(\overline{OB}) = 2\,m\left(\overline{O'D}\right)$.	4. Substitution.

Continued on the next page.

5. $m(\overline{OB}) = m\left(\overline{O'D}\right)$.	5. Equal quantities divided by equal quantities are equal.
6. $\overline{OB} \cong \overline{O'D}$.	6. Segments with equal measure are congruent.
7. Circle O and circle O′ are congruent.	7. If the radii of two circles are congruet then the circles are congruent.

32. Given: Secants $\overline{AC}$ and $\overline{EC}$.

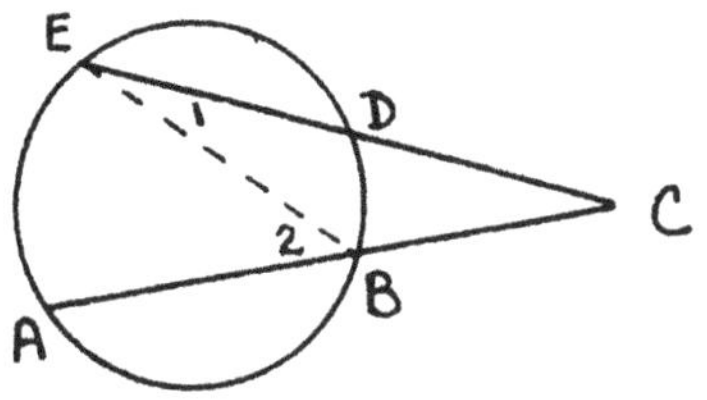

Prove: $m(\angle C) = \frac{1}{2}\left(m(\widehat{AE}) - m(\widehat{BD})\right)$.

STATEMENTS	REASONS
1. Secants $\overline{AC}$ and $\overline{EC}$.	1. Given.
2. $m(\angle 2) = \frac{1}{2}m(\widehat{AE})$. $m(\angle 1) = \frac{1}{2}m(\widehat{BD})$.	2. Theorem 7.3.
3. $m(\angle 2) = m(\angle 1) + m(\angle C)$.	3. Corollary 3.13.
4. $m(\angle C) = m(\angle 2) - m(\angle 1)$.	4. Equal quantities subtracted from equal quantities are equal.
5. $m(\angle C) = \frac{1}{2}m(\widehat{AE}) - \frac{1}{2}m(\widehat{BD})$.	5. Substitution.
6. $m(\angle C) = \frac{1}{2}\left(m(\widehat{AE}) - m(\widehat{BD})\right)$.	6. Distributive law.

Exercise 7.2 (page 254)

16. Given: Circle O and circle O′ are congruent.
$\overline{AB} \cong \overline{CD}$.

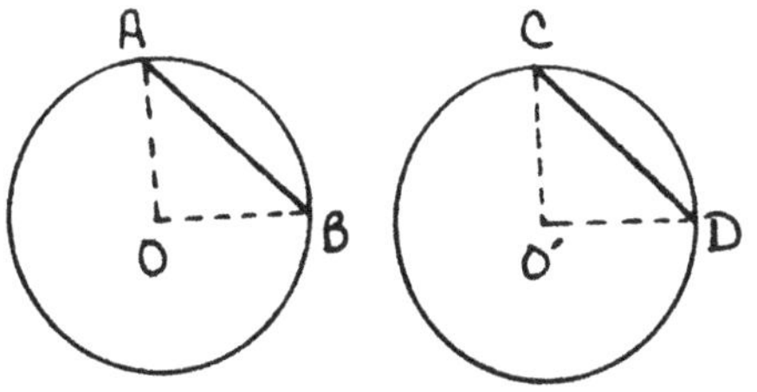

Prove: $\widehat{AB} \cong \widehat{CD}$.

Construction: Draw radii $\overline{OA}$, $\overline{OB}$, $\overline{O'C}$, and $\overline{O'D}$.

The proof is on the next page.

STATEMENTS	REASONS
1. Circle O and circle O′ are congruent.	1. Given.
2. $\overline{AB} \cong \overline{CD}$.	2. Given.
3. Draw $\overline{OA}$, $\overline{OB}$, $\overline{O'C}$, $\overline{O'D}$.	3. Two points determine a line.
4. $\overline{OA} \cong \overline{OB} \cong \overline{O'C} \cong \overline{O'D}$.	4. Radii of congruent circles are congruent.
5. $\triangle AOB \cong \triangle CO'D$.	5. SSS ≅ SSS.
6. $\angle O \cong \angle O'$.	6. cpctc.
7. $\overset{\frown}{AB} \cong \overset{\frown}{CD}$.	7. Congruent central angles have congruent arcs.

18. Given: Circle O and circle O′ are congruent.

$\overline{AB} \cong \overline{CD}$.

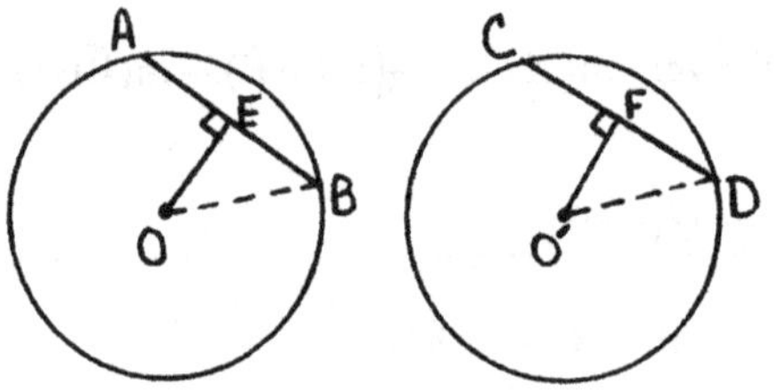

Prove: The distance from each chord and its respective center are equal.

Construction: Draw $\overline{OE} \perp \overline{AB}$ and $\overline{O'F} \perp \overline{CD}$ to form right triangles OEB and O′FD. $\overline{OE}$ and $\overline{O'F}$ represent the distances the chords are from the centers of the circles.

STATEMENTS	REASONS
1. Circle O and circle O′ are congruent; $\overline{AB} \cong \overline{CD}$.	1. Given.
3. $\overline{OE} \perp \overline{AB}$; $\overline{O'F} \perp \overline{CD}$; right triangles OEB and O′FD.	3. By construction.
4. $\overline{AE} \cong \overline{EB}$; $\overline{DF} \cong \overline{FC}$.	4. Theorem 7.7.
5. $\overline{EB} \cong \overline{FD}$.	5. If the segments are congruent, then the segments representing halves of the segments are congruent.
6. $\overline{OB} \cong \overline{O'D}$.	6. Radii of congruent circles are congruent.
7. $\triangle OEB \cong \triangle O'FD$.	7. HL ≅ HL.
8. $\overline{OE} \cong \overline{O'F}$.	8. cpctc.
9. $m(\overline{OE}) = m(\overline{O'F})$.	9. Congruent segments have equal measure.
10. $\overline{AB}$ and $\overline{CD}$ are equidistant from their respective centers.	10. The distances from O and O′ are equal.

Exercise 7.3 (page 258)

18. Given: $\overline{AB}$ and $\overline{DB}$ are tangent to circle O′.

$\overline{DB}$ and $\overline{BC}$ are tangent to circle O.

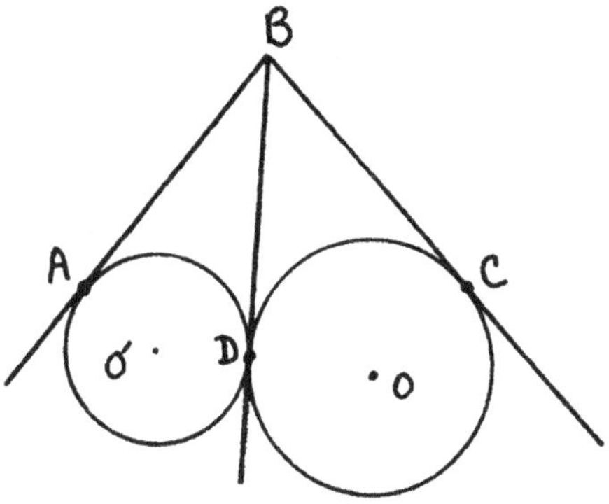

Prove: $\overline{AB} \cong \overline{BC}$.

STATEMENTS	REASONS
1. $\overline{AB}$ and $\overline{DB}$ are tangent to circle O′. $\overline{AB}$ and $\overline{DB}$ are tangent to circle O′.	1. Given.
2. $\overline{AB} \cong \overline{DB}$; $\overline{DB} \cong \overline{BC}$.	2. Theorem 7.14.
3. $\overline{AB} \cong \overline{BC}$.	3. Substitution.

20. Given: $\angle C$ is inscribed in $\widehat{AB}$, which is a semicircle.

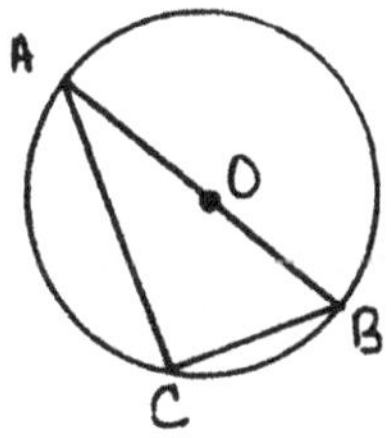

Prove: $m(\angle C) = 90°$.

STATEMENTS	REASONS
1. $\angle C$ is inscribed in $\widehat{AB}$ which is a semicircle.	1. Given.
2. $m(\widehat{AB}) = 180°$.	2. There are 180° in the measure of a semicircle.
3. $m(\angle C) = \frac{1}{2}m(\widehat{AB})$.	3. Theorem 7.2.
4. $m(\angle C) = \frac{1}{2}(180°)$.	4. Substitution.
5. $m(\angle C) = 90°$.	5. Multiplication.

Exercise 7.4 (page 268)

24. Given: In circle O, $\overline{AB} \parallel \overline{CD}$.

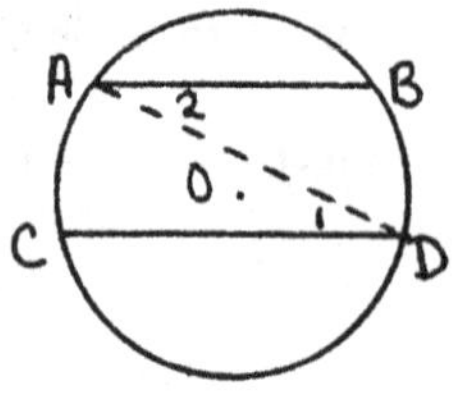

Prove: $\widehat{AC} \cong \widehat{BD}$.

Construction: Draw segment $\overline{AD}$.

STATEMENTS	REASONS
1. In circle O, $\overline{AB} \parallel \overline{CD}$.	1. Given.
2. Draw segment $\overline{AD}$.	2. Two points determine a line.
3. $m(\angle 1) = \frac{1}{2}m(\widehat{AC})$; $m(\angle 2) = \frac{1}{2}m(\widehat{BD})$.	3. Theorem 7.2.
4. $\angle 1 \cong \angle 2$.	4. Theorem 3.5.
5. $m(\angle 1) = m(\angle 2)$.	5. Congruent angles have equal measure.
6. $\frac{1}{2}m(\widehat{AC}) = \frac{1}{2}m(\widehat{BD})$.	6. Substitution.
7. $m(\widehat{AC}) = m(\widehat{BD})$.	7. Equal quantities multiplied by equal quantities are equal.
8. $\widehat{AC} \cong \widehat{BD}$.	8. Arcs with equal measure are congruent.

26. Given: Circles O and O′are congruent.

$\overline{AB}$ and $\overline{CD}$ are common external tangents.

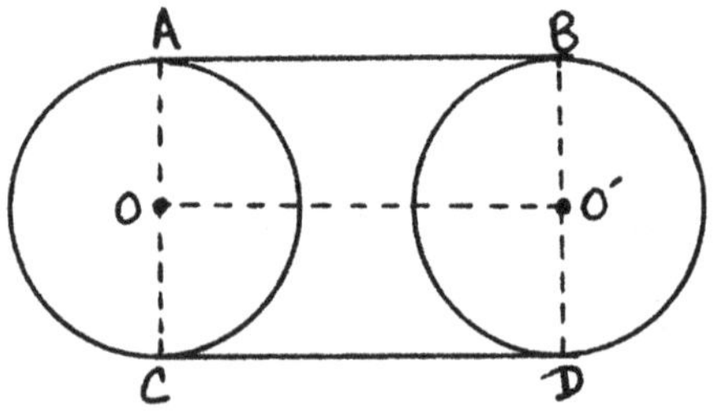

Prove: $\overline{AB} \cong \overline{CD}$.

Construction: Draw radii $\overline{OA}$, $\overline{OC}$, $\overline{O'B}$, and $\overline{O'D}$.

Draw line segment $\overline{OO'}$.

STATEMENTS	REASONS
1. Circles O and O′are congruent.	1. Given.
2. $\overline{AB}$ and $\overline{CD}$ are common external tangents.	2. Given.

Continued on the next page.

3. Draw radii $\overline{OA}$, $\overline{OC}$, $\overline{O'B}$, and $\overline{O'D}$. Draw line segment $\overline{OO'}$.	3. Two points determine a line.
4. $\overline{OA} \perp \overline{AB}$, $\overline{O'B} \perp \overline{AB}$; $\overline{OC} \perp \overline{CD}$, $\overline{O'D} \perp \overline{CD}$.	4. Postulate 7.1
5. $\overline{OA} \parallel \overline{O'B}$; $\overline{OC} \parallel \overline{O'D}$.	5. Lines perpendicular to the same line are parallel.
6. $\overline{OA} \cong \overline{O'B}$; $\overline{OC} \cong \overline{O'D}$.	6. Definition 7.7.
7. OO'BA is a ▱; OO'DC is a ▱.	7. If a pair of opposite sides are parallel and congruent the quadrilateral is a parallelogram.
8. $\overline{AB} \cong \overline{O'O}$; $\overline{CD} \cong \overline{OO'}$.	8. Opposite sides of a parallelogram are congruent.
9. $\overline{AB} \cong \overline{CD}$.	9. Substitution.

Exercise 8.1 (page 281)

16. Given: $a > b$.

$b > c$.

Prove: $a > c$.

STATEMENTS	REASONS
1. $a > b$; $b > c$.	1. Given.
2. $c < b$; $b < a$.	2. Definition 8.1.
3. $c < a$.	3. Theorem 8.6.
4. $a > c$.	4. Definition 8.1.

Exercise 8.2 (page 288)

16. Given: The information in the figure.

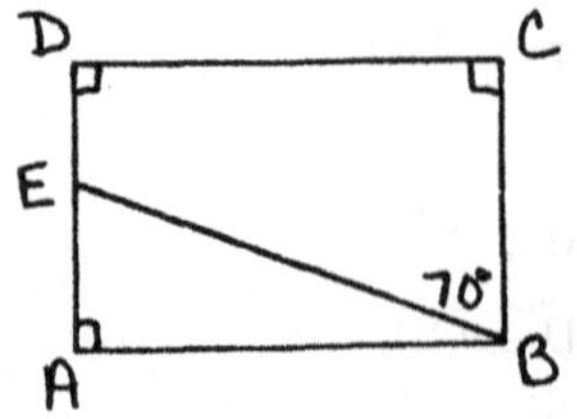

Prove: $m(\overline{AB}) > m(\overline{DA})$.

STATEMENTS	REASONS
1. $\angle A$, $\angle D$, and $\angle C$ are right angles.	1. Given.
2. $m(\angle EBC) = 70^\circ$.	2. Given.
3. $m(\angle A) = m(\angle D) = m(\angle C) = 90^\circ$.	3. Right angles have a measure of 90°.
4. $m(\angle A) + m(\angle B) + m(\angle C) + m(\angle D) = 360^\circ$.	4. The sum of the measure of the angles of a quadrilateral is 360°.
5. $90^\circ + m(\angle B) + 90^\circ + 90^\circ = 360^\circ$.	5. Substitution.
6. $m(\angle B) = 90^\circ$.	6. Equal quantities subtracted from equal quantities are equal.
7. $m(\angle EBA) = 20^\circ$.	7. $\angle EBC$ and $\angle EBA$ are complementary.
8. $m(\angle AEB) = 70^\circ$.	8. $\angle AEB$ and $\angle EBA$ are complementary.
9. $m(\overline{AB}) > m(\overline{AE})$.	9. Theorem 8.10.

18. Given: $m(\angle D) > m(\angle A)$.

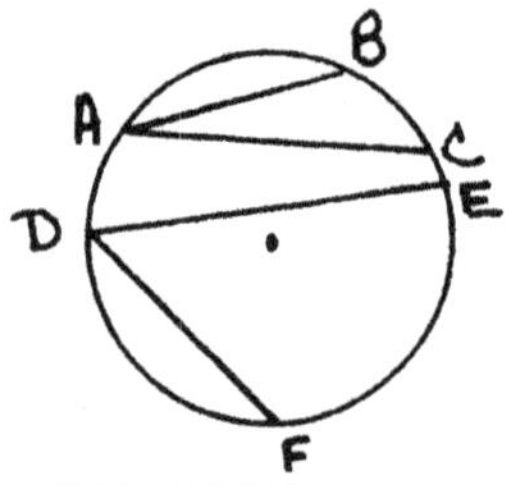

Prove: $m(\widehat{EF}) > m(\widehat{BC})$.

STATEMENTS	REASONS
1. $m(\angle D) > m(\angle A)$.	1. Given.
2. $m(\angle D) = \frac{1}{2}m(\widehat{EF})$; $m(\angle A) = \frac{1}{2}m(\widehat{BC})$.	2. The measure of an inscribed angle is $\frac{1}{2}$ the measure of its intercepted arc.
3. $\frac{1}{2}m(\widehat{EF}) > \frac{1}{2}m(\widehat{BC})$.	3. Substitution. (Postulate 1.9)
4. $m(\widehat{EF}) > m(\widehat{BC})$.	4. Theorem 8.4.

20. Given: $\overline{PR} \perp \overline{RS}$.

Prove: $m(\overline{PS}) > m(\overline{PR})$.

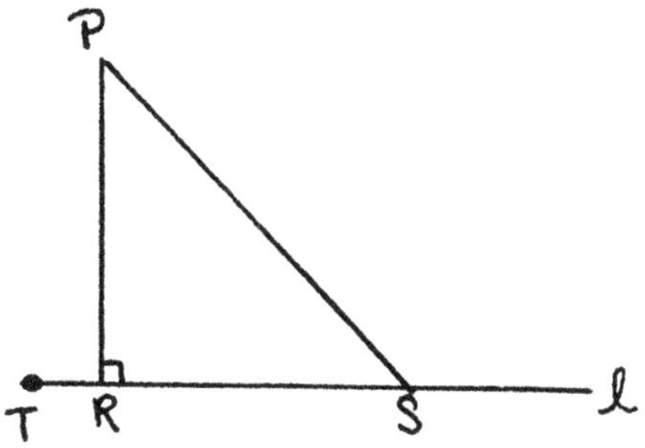

STATEMENTS	REASONS
1. $\overline{PR} \perp \overline{RS}$.	1. Given.
2. $m(\angle PRT) = m(\angle PRS) = 90°$.	2. Perpendicular lines meet and form 90° angles.
3. $m(\angle PRT) > m(\angle PSR)$.	3. Theorem 8.8.
4. $m(\angle PRS) > m(\angle PSR)$.	4. Substitution.
5. $m(\overline{PS}) > m(\overline{PR})$.	5. Theorem 8.10.

22. Given: $m(\overline{AC}) = m(\overline{DF})$.
$m(\overline{BC}) = m(\overline{EF})$.
$m(\angle C) > m(\angle F)$.

Prove: $m(\overline{AB}) > m(\overline{DE})$.

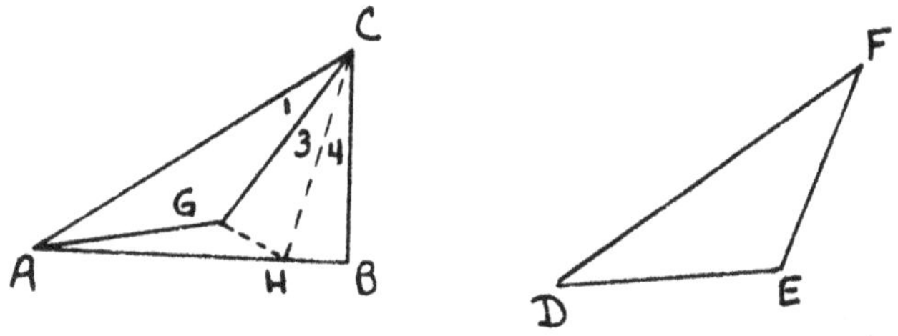

Construction: Construct $\angle 1 \cong \angle F$ and $GC \cong \overline{EF}$. Also construct $\overline{CH}$, the bisector of $\angle GCB$, and draw segment $\overline{GH}$.

STATEMENTS	REASONS
1. $m(\overline{AC}) = m(\overline{DF})$; $m(\overline{BC}) = m(\overline{EF})$; $m(\angle C) > m(\angle F)$.	1. Given.
2. $\angle 1 \cong \angle F$; $\overline{GC} \cong \overline{EF}$.	2. By construction.
3. $m(\angle 1) = m(\angle F)$; $m(\overline{GC}) = m(\overline{EF})$.	3. Congruent angles have equal measure; congruent segments have equal measure.
4. $\triangle ACG \cong \triangle DFE$.	4. SAS $\cong$ SAS.
5. $\overline{AG} \cong \overline{DE}$.	5. cpctc.
6. $m(\overline{AG}) = m(\overline{DE})$.	6. Congruent segments have equal measure.

Continued on the next page.

7. $\overline{CH}$ bisects $\angle GCB$.	7. By construction.
8. $\angle 3 \cong \angle 4$.	8. An angle bisector divides an angle into two congruent angles.
9. $\overline{CH} \cong \overline{CH}$.	9. Reflexive law.
10. $m(\overline{BC}) = m(\overline{GC})$.	10. Substitution.
11. $\overline{BC} \cong \overline{GC}$.	11. Segments with equal measure are congruent.
12. $\triangle GCH \cong \triangle BCH$.	12. SAS $\cong$ SAS.
13. $\overline{HG} \cong \overline{HB}$.	13. cpctc.
14. $m(\overline{HG}) = m(\overline{HB})$.	14. Congruent segments have equal measure.
15. $m(\overline{AH}) + m(\overline{HG}) > m(\overline{AG})$.	15. Postulate 8.2.
16. $m(\overline{AH}) + m(\overline{HB}) > m(\overline{AG})$.	16. Substitution.
17. $m(\overline{AH}) + m(\overline{HB}) = m(\overline{AB})$.	17. Postulate 1.10.
18. $m(\overline{AB}) > m(\overline{DE})$.	18. Substitution.

24. Given: $m(\widehat{AB}) > m(\widehat{CD})$.

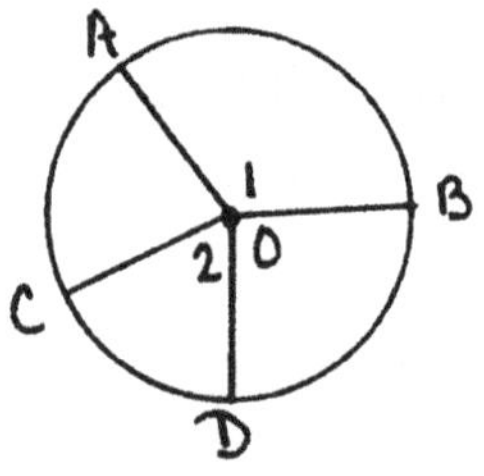

Prove: $m(\angle 1) > m(\angle 2)$.

STATEMENTS	REASONS
1. $m(\widehat{AB}) > m(\widehat{CD})$.	1. Given.
2. $m(\angle 1) = m(\widehat{AB})$; $m(\angle 2) = m(\widehat{CD})$.	2. A central angle has measure equal to its intercepted arc.
3. $m(\angle 1) > m(\angle 2)$.	3. Substitution.

Exercise 8.3 (page 292)

8. Given: Circles O and O′are congruent.

$m(\overline{AB}) > m(\overline{CD})$.

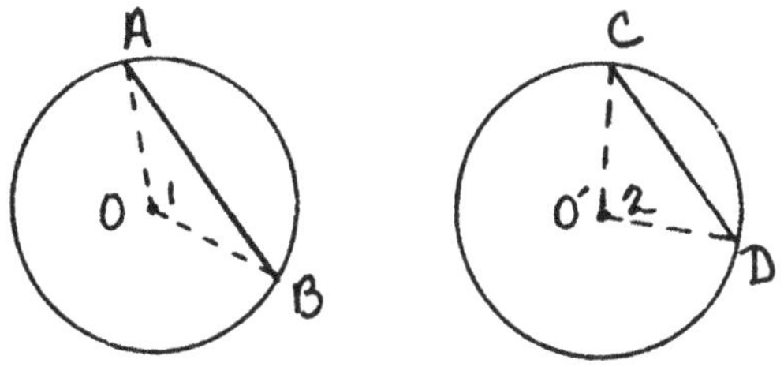

Prove: $m(\overset{\frown}{AB}) > m(\overset{\frown}{CD})$.

Construction: Draw radii $\overline{OA}$, $\overline{OB}$, $\overline{O'C}$, and $\overline{O'D}$.

STATEMENTS	REASONS
1. Circles O and O′are congruent.	1. Given.
2. $m(\overline{AB}) > m(\overline{CD})$.	2. Given.
3. Draw radii $\overline{OA}$, $\overline{OB}$, $\overline{O'C}$, and $\overline{O'D}$.	3. Two points determine a line.
4. $m(\overline{OA}) = m(\overline{OB}) = m(\overline{O'C}) = m(\overline{O'D})$.	4. All radii in congruent circles have equal measure.
5. $m(\angle 1) > m(\angle 2)$.	5. Theorem 8.12.
6. $m(\overset{\frown}{AB}) > m(\overset{\frown}{CD})$.	6. Theorem 8.13.

10. Given: $m(\overline{CD}) < m(\overline{AB})$.

$\overline{OE} \perp \overline{CD}$.

$\overline{OF} \perp \overline{AB}$.

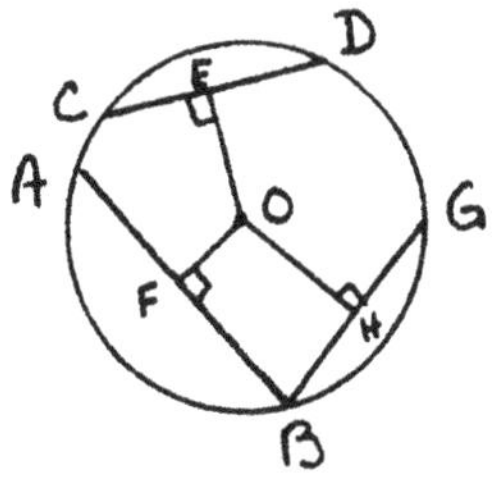

Prove: $m(\overline{OE}) > m(\overline{OF})$.

Construction: Construct $\overline{BG} \cong \overline{CD}$ and $\overline{OH} \perp \overline{BG}$.

STATEMENTS	REASONS
1. $m(\overline{CD}) < m(\overline{AB})$; $\overline{OE} \perp \overline{CD}$; $\overline{OF} \perp \overline{AB}$.	1. Given.
2. $m(\overline{BG}) = m(\overline{CD})$; $\overline{OH} \perp \overline{BG}$.	2. By construction.

Continued on the next page.

3. $m(\overline{OH}) = m(\overline{OE})$.	3. Theorem 7.13.
4. $m(\overline{BG}) < m(\overline{AB})$.	4. Substitution.
5. $m(\overline{OH}) > m(\overline{OF})$.	5. Theorem 8.17.
6. $m(\overline{OE}) > m(\overline{OF})$.	6. Substitution.

12. Given: Point O is on the exterior of $\angle$BAC.

$m(\overline{AB}) > m(\overline{AC})$.

$\overline{OD} \perp \overline{AB}$.

$\overline{OE} \perp \overline{AC}$.

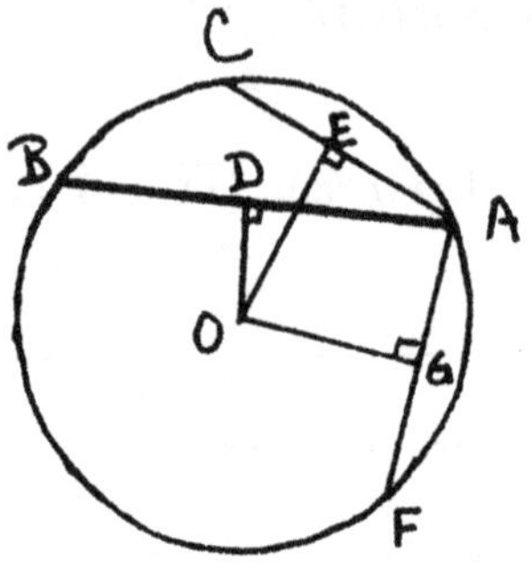

Prove: $m(\overline{OE}) > m(\overline{OD})$.

Construction: Construct $\overline{AF}$ so that O is on the interior of $\angle$BAF, $\overline{AF} \cong \overline{AC}$, and $\overline{OG} \perp \overline{AF}$.

STATEMENTS	REASONS
1. Point O is on the exterior of $\angle$BAC. $m(\overline{AB}) > m(\overline{AC})$; $\overline{OD} \perp \overline{AB}$; $\overline{OE} \perp \overline{AC}$.	1. Given.
2. Point O is on the interior of $\angle$BAF, $m(\overline{AF}) = m(\overline{AC})$; $\overline{OG} \perp \overline{AF}$.	2. By construction.
3. $m(\overline{OG}) = m(\overline{OE})$.	3. Theorem 7.13.
4. $m(\overline{AB}) > m(\overline{AF})$.	4. Substitution.
5. $m(\overline{OG}) > m(\overline{OD})$.	5. Theorem 8.17, case I.
6. $m(\overline{OE}) > m(\overline{OD})$.	6. Substitution.

Exercise 9.1 (page 304)

18. Given: $\triangle$ABC.

Circle O, with diameter $\overline{AB}$.

Point C is on circle O.

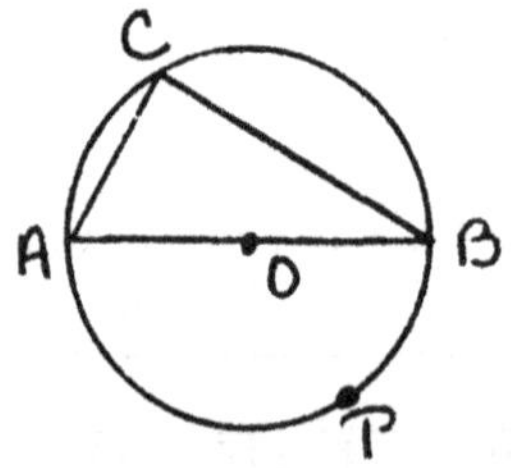

Prove: $\angle$ACB is a right angle.

The proof is on the next page.

STATEMENTS	REASONS
1. $\triangle ABC$; Circle O with diameter $\overline{AB}$; point C is on circle O.	1. Given.
2. $\widehat{APB}$ is a semicircle.	2. If the endpoints of an arc are on the diameter, then the arc is a semicircle.
3. $m(\widehat{APB}) = 180°$.	3. The measure of a semicircle is 180°.
4. $m(\angle ACB) = 90°$.	4. The measure of an inscribed angle is $\frac{1}{2}$ the measure of its intercepted arc.
5. $\angle ACB$ is a right angle.	5. If an angle has measure of 90°, then the angle is a right angle.

Exercise 9.2 (page 311)

16. Given: $\triangle ABC$ with altitudes $\overline{AE}$, $\overline{CD}$, and $\overline{BF}$.

Prove: $\overline{AE}$, $\overline{CD}$, and $\overline{BF}$ meet in a point.

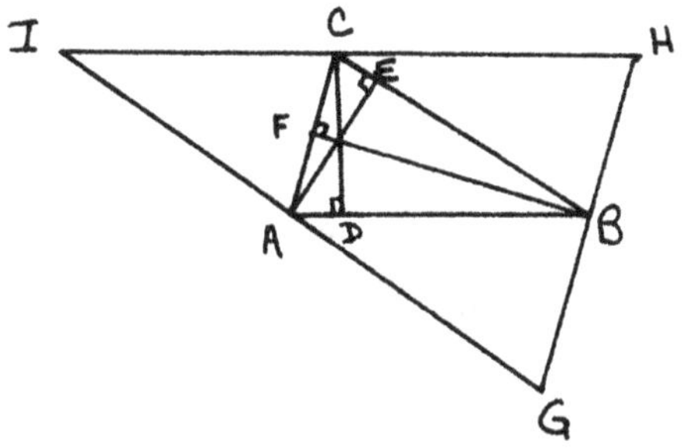

Plan: Construct $\overline{IG} \parallel \overline{CB}$, $\overline{IH} \parallel \overline{AB}$, and $\overline{GH} \parallel \overline{AC}$, and show that $\overline{AE}$, $\overline{CD}$, and $\overline{BF}$ are the perpendicular bisectors of the sides of $\triangle IGH$. They would then be concurrent by Theorem 9.5. To do so, show that $\overline{AE}$, $\overline{CD}$, and $\overline{BF}$ are perpendicular to $\overline{IG}$, $\overline{IH}$, and $\overline{GH}$ respectively. Then show point A is the midpoint of $\overline{IG}$ by considering ▱ABCI and ▱AGBC. In a similar fashion, points B and C can be shown to the the midpoints of $\overline{GH}$ and $\overline{IH}$ respectively.

STATEMENTS	REASONS
1. $\triangle ABC$ with altitudes $\overline{AE}$, $\overline{CD}$, and $\overline{BF}$.	1. Given.
2. $\overline{IG} \parallel \overline{CB}$, $\overline{IH} \parallel \overline{AB}$, and $\overline{GH} \parallel \overline{AC}$.	2. By construction.
3. ICBA and CBGA are parallelograms.	3. If opposite sides of a quadrilateral are parallel, then the quadrilateral is a paralllelogram.
4. $\overline{IA} \cong \overline{CB}$; $\overline{CB} \cong \overline{AG}$.	4. Opposite sides of a parallelogram are congruent.

Continued on the next page.

5. $\overline{IA} \cong \overline{AG}$.	5. Substitution.
6. A is the midpoint of $\overline{IG}$.	6. If a point divides a segment into two congruent segments, it is a midpoint of the segment.
7. $\overline{AE} \perp \overline{CB}$; $\overline{CD} \perp \overline{AB}$; $\overline{BF} \perp \overline{AC}$.	7. Altitudes of a triangle are perpendicular to the opposite side.
8. $\overline{AE} \perp \overline{IG}$.	8. If a line is perpendicular to one of two parallel lines, it is perpendicular to the other.
9. $\overline{AE}$ is the perpendicular bisector of $\overline{IG}$.	9. Definition of a perpendicular bisector.
10. Similarly, $\overline{AE}$ is the perpendicular bisector of $\overline{IG}$. $\overline{AE}$ is the perpendicular bisector of $\overline{IG}$.	10. Reasons 3 – 8.
11. $\overline{AE}$, $\overline{CD}$, and $\overline{BF}$ meet in a point.	11. Theorem 9.5.

18. Given: $\triangle ABC$ with medians $\overline{AN}$, $\overline{BM}$, and $\overline{CQ}$.

Prove: $m(\overline{AP}) = \frac{2}{3}m(\overline{AN})$; $m(\overline{BP}) = \frac{2}{3}m(\overline{BM})$, and $m(\overline{CP}) = \frac{2}{3}m(\overline{CQ})$.

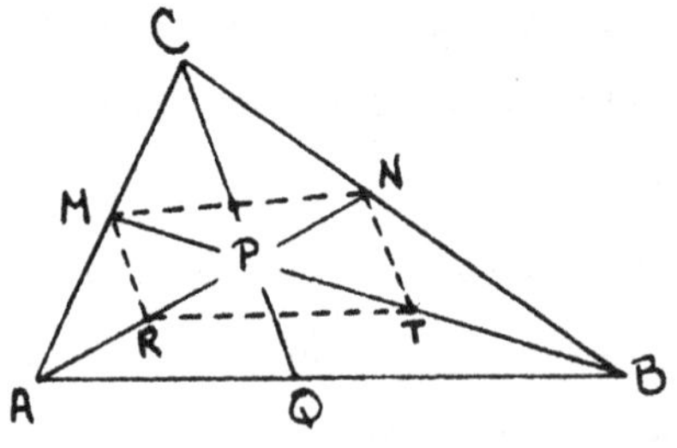

Plan: Locate the midpoints of segments $\overline{AP}$ and $\overline{BP}$ and call them R and T, respectively. Draw segments $\overline{MR}$, $\overline{RT}$, $\overline{MN}$, and $\overline{TN}$. Show that MRTN is a parallelogram and that $m(\overline{RP}) = m(\overline{PN})$ and $m(\overline{TP}) = m(\overline{PM})$. Therefore, $m(\overline{AR}) = m(\overline{RP}) = m(\overline{PN})$, and $m(\overline{BT}) = m(\overline{TP}) = m(\overline{PM})$ and $m(\overline{AP}) = \frac{2}{3}m(\overline{AN})$ and $m(\overline{BP}) = \frac{2}{3}m(\overline{BM})$. To show $m(\overline{CP}) = \frac{2}{3}m(\overline{CQ})$, consider the midpoints of segments $\overline{PC}$ and $\overline{AP}$ and proceed with a similar argument.

The proof is on the next page.

Construction: Construct R and T as midpoints of $\overline{AP}$ and $\overline{BP}$. Draw segments $\overline{MR}$, $\overline{RT}$, $\overline{MN}$, and $\overline{TN}$.

STATEMENTS	REASONS
1. $\triangle ABC$ with medians $\overline{AN}$, $\overline{BM}$, and $\overline{CQ}$.	1. Given.
2. R is the midpoint of $\overline{AP}$; T is the midpoint of $\overline{BP}$.	2. By construction.
3. $\overline{MN} \parallel \overline{AB}$; $m(\overline{MN}) = \frac{1}{2}m(\overline{AB})$.	3. Theorem 4.17.
4. $\overline{RT} \parallel \overline{AB}$; $m(\overline{RT}) = \frac{1}{2}m(\overline{AB})$.	4. Theorem 4.17.
5. $\overline{MN} \parallel \overline{RT}$.	5. Theorem 3.23.
6. $m(\overline{MN}) = m(\overline{RT})$.	6. Substitution.
7. $\overline{MN} \cong \overline{RT}$.	7. Segments with equal measure are congruent.
8. MNTR is a parallelogram.	8. Theorem 4.7.
9. $\overline{RP} \cong \overline{PN}$; $\overline{TP} \cong \overline{PM}$.	9. Theorem 4.8.
10. $\overline{AR} \cong \overline{RP}$; $\overline{BT} \cong \overline{TP}$	10. A midpoint divides a segment into congruent segments.
11. $\overline{AR} \cong \overline{RP} \cong \overline{PN}$; $\overline{BT} \cong \overline{TP} \cong \overline{PM}$.	11. Segments congruent to congruent segments are congruent to each other.
12. $m(\overline{AR}) = m(\overline{RP}) = m(\overline{PN})$; $m(\overline{BT}) = m(\overline{TP}) = m(\overline{PM})$.	12. Congruent segments have equal measure.
13. $m(\overline{AN}) = m(\overline{AR}) + m(\overline{RP}) + m(\overline{PN})$.	13. Postulate 1.10.
14. $m(\overline{AN}) = m(\overline{AR}) + m(\overline{AR}) + m(\overline{AR})$.	14. Substitution.
15. $m(\overline{AN}) = 3\,m(\overline{AR})$.	15. Substitution.
16. $m(\overline{AR}) = \frac{1}{3}m(\overline{AN})$.	16. Equal quantities divided by equal quantities are equal.
17. $m(\overline{AP}) = m(\overline{AR}) + m(\overline{RP})$.	17. Postulate 1.10.
18. $m(\overline{AP}) = m(\overline{AR}) + m(\overline{AR})$.	18. Substitution.
19. $m(\overline{AP}) = 2\,m(\overline{AR})$.	19. Substitution.
20. $m(\overline{AP}) = 2\left(\frac{1}{3}m(\overline{AN})\right)$.	20. Substitution.
21. $m(\overline{AP}) = \frac{2}{3}m(\overline{AN})$.	21. Substitution.

A similar argument will show that $m(\overline{BP}) = \frac{2}{3}m(\overline{BM})$, and that $m(\overline{CP}) = \frac{2}{3}m(\overline{CQ})$.

Exercise 10.1 (page 322)

16. Given: Regular polygon ABCDE with radii $\overline{OA}$, $\overline{OB}$, $\overline{OC}$, $\overline{OD}$, and $\overline{OE}$.

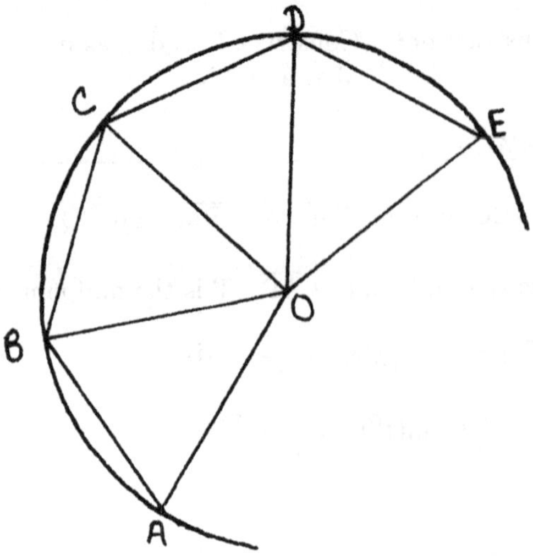

Prove: $\overline{OA} \cong \overline{OB} \cong \overline{OC} \cong \overline{OD} \cong \overline{OE}$.

Construction: Draw the circumscribed circle about the polygon.

STATEMENTS	REASONS
1. Regular polygon ABCDE with radii $\overline{OA}$, $\overline{OB}$, $\overline{OC}$, $\overline{OD}$, and $\overline{OE}$.	1. Given.
2. Circle O is circumscribed about the regular polygon.	2. Theorem 10.3.
3. $\overline{OA} \cong \overline{OB} \cong \overline{OC} \cong \overline{OD} \cong \overline{OE}$.	3. Radii of the same circle are congruent.

18. Given: Regular polygon ABCDE with radius $\overline{OC}$.

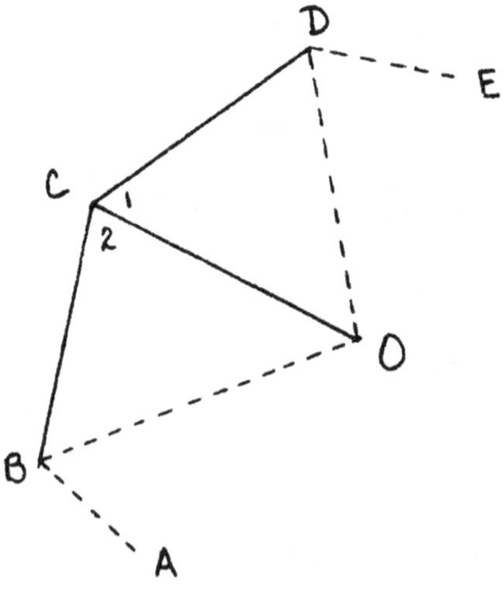

Prove: $\angle 1 \cong \angle 2$.

Construction: Draw radii $\overline{OB}$ and $\overline{OD}$.

STATEMENTS	REASONS
1. Regular polygon ABCDE with radius $\overline{OC}$.	1. Given.
2. Draw radii $\overline{OB}$ and $\overline{OD}$.	2. Two points determine a line.
3. $\overline{OD} \cong \overline{OB}$.	3. Radii of the same regular polygon are congruent.
4. $\overline{OC} \cong \overline{OC}$.	4. Reflexive law.

Continued on the next page.

5. $\overline{CB} \cong \overline{CD}$.	5. All sides of a regular polygon are congruent.
6. $\triangle BOC \cong \triangle COD$.	6. SSS ≅ SSS.
7. $\angle 1 \cong \angle 2$.	7. cpctc.

Exercise 10.2 (page 329)

18. Given: Regular polygons ABCD. . . and FGHI. . . with $\overline{OE}$ and $\overline{O'J}$ as respective apothems.

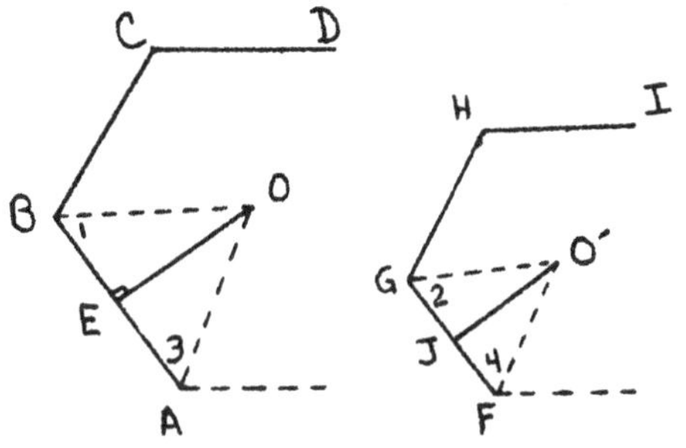

Prove: $\dfrac{P_{ABCD...}}{P_{FGHI...}} = \dfrac{m(\overline{OE})}{m(\overline{O'J})}$.

Construction: Draw radii $\overline{OA}$, $\overline{OB}$, $\overline{O'F}$, and $\overline{O'G}$.

STATEMENTS	REASONS
1. Regular polygons ABCD. . . and FGHI. . . .	1. Given.
2. $\dfrac{m(\overline{AB})}{m(\overline{FG})} = \dfrac{m(\overline{BC})}{m(\overline{GH})} = \dfrac{m(\overline{CD})}{m(\overline{HI})} = \ldots$.	2. Theorem 6.6.
3. $\dfrac{m(\overline{AB}) + m(\overline{BC}) + m(\overline{CD}) + \ldots}{m(\overline{FG}) + m(\overline{GH}) + m(\overline{HI}) + \ldots} = \dfrac{m(\overline{AB})}{m(\overline{GF})}$.	3. Theorem 6.8.
4. $m(\overline{AB}) + m(\overline{BC}) + m(\overline{CD}) + \ldots = P_{ABCD...}$; $m(\overline{GF}) + m(\overline{GH}) + m(\overline{HI}) + \ldots = P_{FGHI...}$.	4. Definition of perimeter.
5. $\dfrac{P_{ABCD...}}{P_{FGHI...}} = \dfrac{m(\overline{AB})}{m(\overline{GF})}$.	5. Substitution.
6. $\angle CBA \cong \angle HGF$.	6. Definition 6.6.

Continued on the next page.

STATEMENTS	REASONS
7. $m(\angle 1) = \frac{1}{2}m(\angle CBA)$; $m(\angle 2) = \frac{1}{2}m(\angle HGI)$.	7. Exercise 18, section 10.1.
8. $m(\angle 1) = m(\angle 2)$.	8. Halves of equals are equal.
9. $\angle 1 \cong \angle 2$.	9. Angles with equal measure are congruent.
10. $\overline{OB} \cong \overline{OA}$; $\overline{O'G} \cong \overline{O'F}$	10. All radii in a regular polygon are congruent.
11. $\angle 1 \cong \angle 3$; $\angle 2 \cong \angle 4$.	11. Theorem 2.2.
12. $\angle 3 \cong \angle 4$.	12. Substitution.
13. $\triangle AOB \sim \triangle FO'G$.	13. Postulate 6.1.
14. $\frac{m(\overline{OE})}{m(\overline{O'J})} = \frac{m(\overline{AB})}{m(\overline{GF})}$.	14. Exercise 36 in section 6.3.
15. $\frac{P_{ABCD...}}{P_{FGHI...}} = \frac{m(\overline{OE})}{m(\overline{O'J})}$.	15. Substitution.

20. Given: ABCDE is a regular polygon.

$\overline{OM}$ is an apothem.

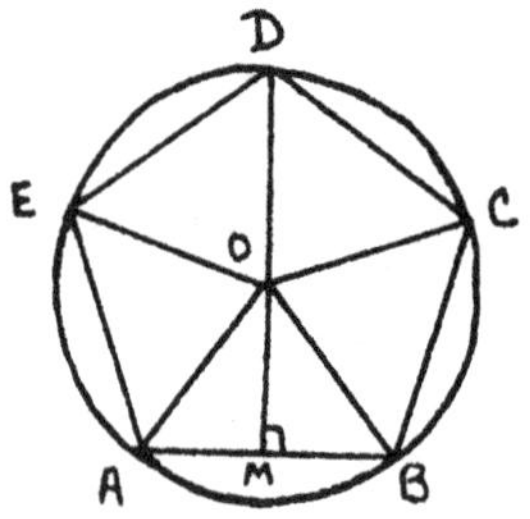

Prove: $A(ABCDE) = \frac{1}{2}m(\overline{OM})\,P_{ABCDE}$.

Proof: We find the areas of $\triangle ABO$, $\triangle BCO$, $\triangle CDO$, $\triangle DEO$, and $\triangle EAO$ and add them together to find the area of the regular polygon ABCDE.

STATEMENTS	REASONS
1. ABCDE is a regular polygon with apothem $\overline{OM}$.	1. Given.
2. $\overline{OA} \cong \overline{OB} \cong \overline{OC} \cong \overline{OD} \cong \overline{OE}$.	2. All radii of a regular polygon are congruent.
3. $\overline{AB} \cong \overline{BC} \cong \overline{CD} \cong \overline{DE} \cong \overline{EA}$.	3. All sides of a regular polygon are congruent.

Continued on the next page.

4. $\triangle AOB \cong \triangle BOC \cong \triangle COD \cong \triangle DOE \cong \triangle EOA$. 4. SSS ≅ SSS.

5. $A(ABCDE) = A(\triangle AOB + \triangle BOC + \triangle COD + \triangle DOE + \triangle EOA)$. 5. Postulate 5.4.

6. The altitude of each triangle is $\overline{OM}$. 6. Exercise 17 in section 10.2.

7. $A(ABCDE) = \frac{1}{2}m(\overline{OM})\,m(\overline{AB}) + \frac{1}{2}m(\overline{OM})\,m(\overline{BC}) + \frac{1}{2}m(\overline{OM})\,m(\overline{CD}) + \frac{1}{2}m(\overline{OM})\,m(\overline{DE}) + \frac{1}{2}m(\overline{OM})\,m(\overline{EA})$. 7. Theorem 5.2.

8. $A(ABCDE) = \frac{1}{2}m(\overline{OM})\Big(m(\overline{AB}) + m(\overline{BC}) + m(\overline{CD}) + m(\overline{DE}) + m(\overline{EA})\Big)$. 8. Distributive law.

9. $P = m(\overline{AB}) + m(\overline{BC}) + m(\overline{CD}) + m(\overline{DE}) + m(\overline{EA})$. 9. Definition of perimeter.

10. $A(ABCDE) = \frac{1}{2}m(\overline{OM})\,P_{ABCDE}$. 10. Substitution.

Chapter Ten Review Exercises (page 330)

20. Given: $\triangle ABC$ is a right triangle with $\angle C$ a right angle.

The sides of the right triangle ABC are diameters of the semicircles shown.

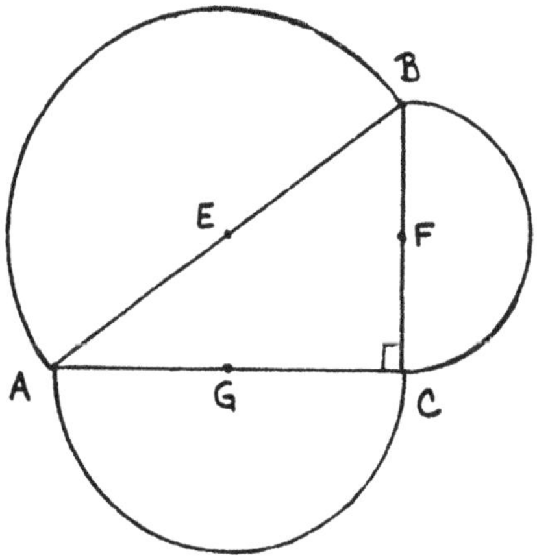

Prove: The sum of the areas of the semicircles drawn on the triangle's legs is equal to the area of the semicircle drawn on the hypotenuse.

Construction: Construct the centers of the semicircles and label them E, F, and G.

STATEMENTS	REASONS
1. $\triangle ABC$ is a right triangle with $\angle C$ a right angle.	1. Given.
2. $m(\overline{AG}) = \frac{1}{2}m(\overline{AC})$; $m(\overline{BF}) = \frac{1}{2}m(\overline{BC})$; $m(\overline{AE}) = \frac{1}{2}m(\overline{AB})$.	2. The radius is equal to one-half the measure of the diameter.
3. $2\,m(\overline{AG}) = m(\overline{AC})$; $2\,m(\overline{BF}) = m(\overline{BC})$; $2\,m(\overline{AE}) = m(\overline{AB})$.	3. Equal quantities multiplied by equal quantities are equal.

Continued on the next page.

4.	$A(\text{circle G}) = \pi[m(\overline{AG})]^2$; $A(\text{circle F}) = \pi[m(\overline{BF})]^2$; $A(\text{circle E}) = \pi[m(\overline{AE})]^2$.	4.	Definition of area of a circle.
5.	$A(\text{semicircle G}) = \frac{1}{2}\pi[m(\overline{AG})]^2$; $A(\text{semicircle F}) = \frac{1}{2}\pi[m(\overline{BF})]^2$; $A(\text{semicircle E}) = \frac{1}{2}\pi[m(\overline{AE})]^2$.	5.	The area of a semicircle is one-half the area of the circle.
6.	$[m(\overline{AC})]^2 + [m(\overline{BC})]^2 = [m(\overline{AB})]^2$.	6.	Pythagorean theorem.
7.	$[2\,m(\overline{AG})]^2 + [2\,m(\overline{BF})]^2 = [2\,m(\overline{AE})]^2$.	7.	Substitution.
8.	$4\,[m(\overline{AG})]^2 + 4\,[m(\overline{BF})]^2 = 4\,[m(\overline{AE})]^2$.	8.	Substitution. $2^2 = 4$.
9.	$\frac{1}{2}\pi\,[m(\overline{AG})]^2 + \frac{1}{2}\pi\,[m(\overline{BF})]^2 = \frac{1}{2}\pi\,[m(\overline{AE})]^2$.	9.	Equal quantities multiplied by equal quantities are equal. (Multiplied by $\frac{\pi}{8}$).
10.	A(semicircle G) + A(semicircle F) = A(semicircle E).	10.	Substitution.

Exercise 12.1 (page 370)

60. Find the lengths of each side of the triangle whose verticies are at (– 1, 2), (3, 1), and (4, 5).

(– 1, 2); (3, 1)

$$d_1 = \sqrt{(3-(-1))^2 + (1-2)^2} = \sqrt{(4)^2 + (-1)^2} = \sqrt{16+1} = \sqrt{17}$$

(– 1, 2); (4, 5)

$$d_2 = \sqrt{(4-(-1))^2 + (5-2)^2} = \sqrt{(5)^2 + (3)^2} = \sqrt{25+9} = \sqrt{34}$$

(3, 1); (4, 5)

$$d_3 = \sqrt{(4-3)^2 + (5-1)^2} = \sqrt{(1)^2 + (4)^2} = \sqrt{1+16} = \sqrt{17}$$

Since $d_1 = d_3$, the triangle is isosceles.

Exercise 12.4 (page 394)

2.

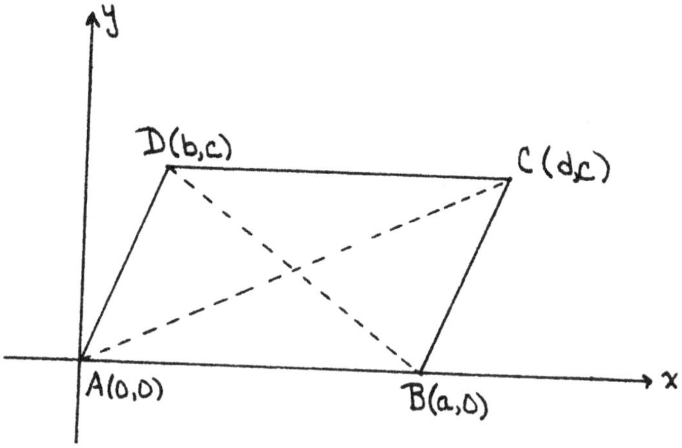

Since ABCD is a parallelogram, we can assign the coordinates as shown in the figure. We can then draw diagonals $\overline{AC}$ and $\overline{BD}$. The midpoint of $\overline{AC}$ has coordinates of $\left(\frac{d}{2}, \frac{c}{2}\right)$ and the midpoint of $\overline{BD}$ has coordinates of $\left(\frac{a+b}{2}, \frac{c}{2}\right)$. Because ABCD is a parallelogram, its opposite sides are congruent. Thus,

$$a - 0 = d - b$$

or $$a + b = d.$$

Because $\left(\frac{a+b}{2}, \frac{c}{2}\right) = \left(\frac{d}{2}, \frac{c}{2}\right)$, the midpoint of $\overline{BD}$ is also the midpoint of $\overline{AC}$. Thus, the diagonals bisect each other.

4.

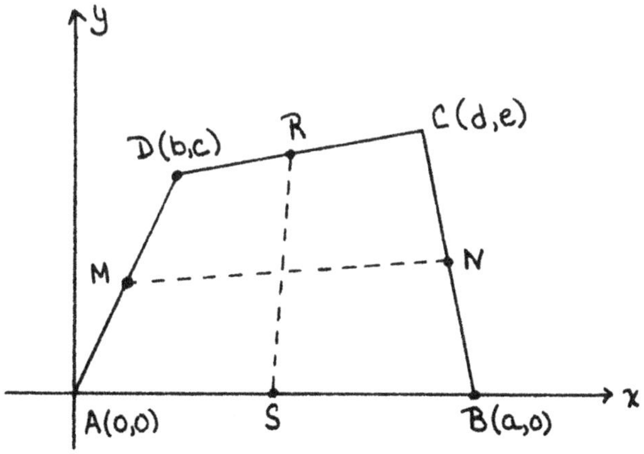

The proof is on the next page.

Since ABCD is a quadrilateral, we can assign coordinates as shown in the figure. Because M, N, R and S are midpoints, their coordinates are $M\left(\frac{b}{2}, \frac{c}{2}\right)$, $N\left(\frac{a+d}{2}, \frac{e}{2}\right)$, $R\left(\frac{b+d}{2}, \frac{e+c}{2}\right)$, and $S\left(\frac{a}{2}, 0\right)$.

$$\text{The midpoint of } \overline{MN} \text{ is } \left(\frac{\frac{a+d}{2}+\frac{b}{2}}{2}, \frac{\frac{c}{2}+\frac{e}{2}}{2}\right) = \left(\frac{a+b+d}{4}, \frac{c+e}{4}\right).$$

$$\text{The midpoint of } \overline{RS} \text{ is } \left(\frac{\frac{a}{2}+\frac{b+d}{2}}{2}, \frac{\frac{c+e}{2}}{2}\right) = \left(\frac{a+b+d}{4}, \frac{c+e}{4}\right).$$

Because the midpoint of $\overline{MN}$ is also the midpoint of $\overline{RS}$, the segments bisect each other.

6.

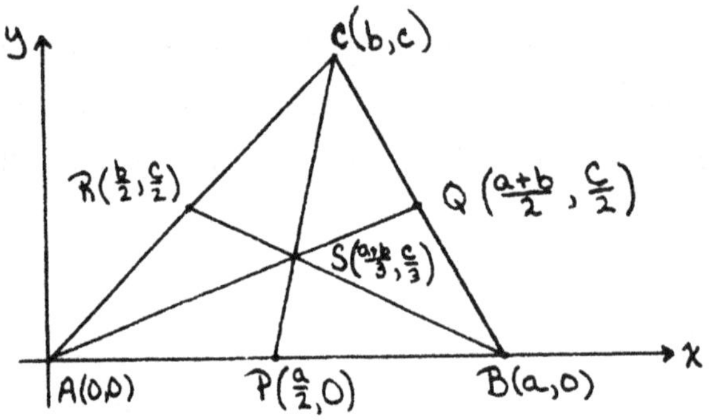

From Figure 12 – 18 and the analytic proof of Theorem 9.8 we have the above coordinates. To prove the point of intersection of the medians of a triangle is two-thirds of the distance from each vertex to the opposite side, use the distance formula.

$$d(\overline{AQ}) = \sqrt{\left(\frac{a+b}{2} - 0\right)^2 + \left(\frac{c}{2} - 0\right)^2} = \sqrt{\frac{a^2+2ab+b^2}{4} + \frac{c^2}{4}} = \frac{\sqrt{a^2+2ab+b^2+c^2}}{2}$$

$$d(\overline{AS}) = \sqrt{\left(\frac{a+b}{3} - 0\right)^2 + \left(\frac{c}{3} - 0\right)^2} = \sqrt{\frac{a^2+2ab+b^2}{9} + \frac{c^2}{9}} = \frac{\sqrt{a^2+2ab+b^2+c^2}}{3}$$

$$\frac{2}{3}\left(d(\overline{AQ})\right) = \frac{2}{3}\left(\frac{\sqrt{a^2+2ab+b^2+c^2}}{2}\right) = \frac{\sqrt{a^2+2ab+b^2+c^2}}{3} = d(\overline{AS})$$

A similar argument will show that $m(\overline{BS}) = \frac{2}{3}m(\overline{BR})$, and that $m(\overline{CS}) = \frac{2}{3}m(\overline{CP})$.

8.

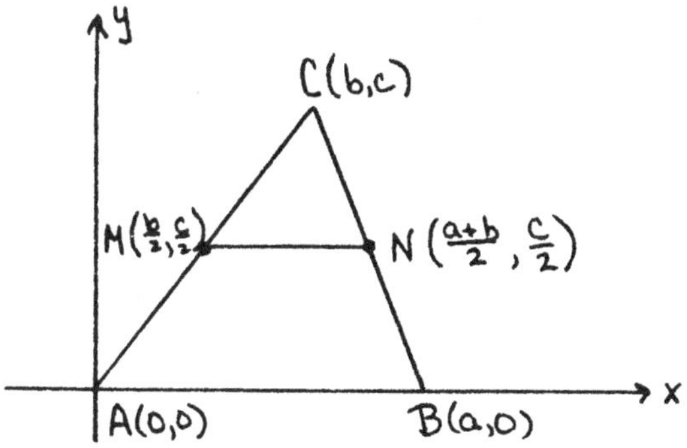

$$d(\overline{MN}) = \sqrt{\left(\frac{a+b}{2} - \frac{b}{2}\right)^2 + \left(\frac{c}{2} - \frac{c}{2}\right)^2} = \sqrt{\left(\frac{a}{2}\right)^2 + (0)^2} = \sqrt{\frac{a^2}{4}} = \frac{a}{2} \qquad (a > 0).$$

$$d(\overline{AB}) = |a - 0| = a \qquad (a > 0).$$

Since $\frac{1}{2}[d(\overline{AB})] = \frac{1}{2}a = \frac{a}{2}$, $d(\overline{MN}) = \frac{1}{2}[d(\overline{AB})]$.

10.

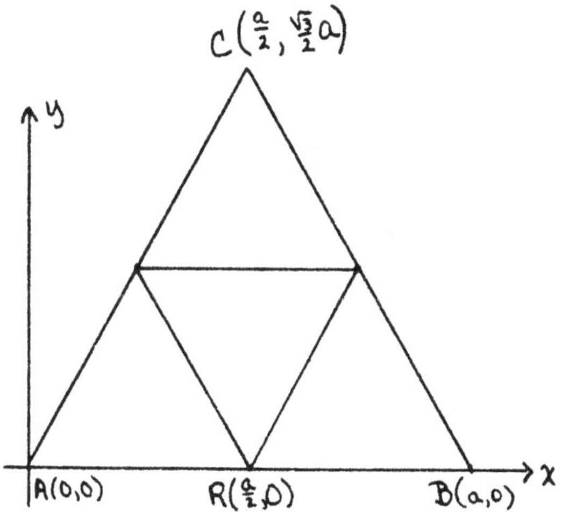

Since $\triangle ABC$ is equilateral, point C's x-coordinate is $\frac{a}{2}$ and the y-coordinate is $\frac{\sqrt{3}\,a}{2}$.
The midpoints of the sides are:

R, the midpoint of $\overline{AB}$ is $\left(\frac{a+0}{2}, \frac{0+0}{2}\right) = \left(\frac{a}{2}, 0\right)$.

S, the midpoint of $\overline{BC}$ is $\left(\frac{\frac{a}{2}+a}{2}, \frac{\frac{\sqrt{3}a}{2}+0}{2}\right) = \left(\frac{3a}{4}, \frac{\sqrt{3}a}{4}\right)$.

T, the midpoint of $\overline{AC}$ is $\left(\frac{\frac{a}{2}+0}{2}, \frac{\frac{\sqrt{3}a}{2}+0}{2}\right) = \left(\frac{a}{4}, \frac{\sqrt{3}a}{4}\right)$.

Continued on the next page.

Finding the distance between each pair of midpoints we have,

$$d(\overline{RS}) = \sqrt{\left(\frac{3a}{4} - \frac{a}{2}\right)^2 + \left(\frac{\sqrt{3}a}{4} - 0\right)^2} = \sqrt{\left(\frac{a}{4}\right)^2 + \left(\frac{\sqrt{3}a}{4}\right)^2} = \sqrt{\frac{a^2}{16} + \frac{3a^2}{16}} = \sqrt{\frac{4a^2}{16}} = \frac{2a}{4} = \frac{a}{2}\ ;\ a > 0$$

$$d(\overline{ST}) = \sqrt{\left(\frac{a}{4} - \frac{3a}{4}\right)^2 + \left(\frac{\sqrt{3}a}{4} - \frac{\sqrt{3}a}{4}\right)^2} = \sqrt{\left(\frac{-2a}{4}\right)^2 + 0^2} = \sqrt{\frac{4a^2}{16}} = \frac{2a}{4} = \frac{a}{2}\ ;\ a > 0$$

$$d(\overline{RT}) = \sqrt{\left(\frac{a}{4} - \frac{a}{2}\right)^2 + \left(\frac{\sqrt{3}a}{4} - 0\right)^2} = \sqrt{\left(-\frac{a}{4}\right)^2 + \left(\frac{\sqrt{3}a}{4}\right)^2} = \sqrt{\frac{a^2}{16} + \frac{3a^2}{16}} = \sqrt{\frac{4a^2}{16}} = \frac{2a}{4} = \frac{a}{2}\ ;\ a > 0$$

Since all three distances are equal, the triangle formed is an equilateral triangle.

12.

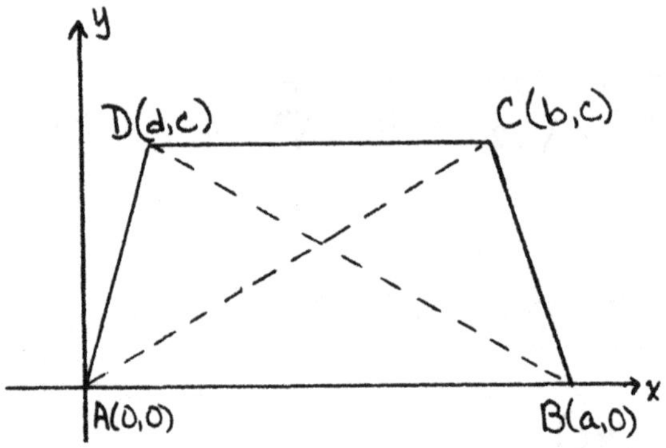

Since ABCD is a trapezoid, we are justified using the coordinates in the figure. Since $\overline{AB} \parallel \overline{DC}$ we give the point D and C the same y-coordinate. We are given that the diagonals are congruent hence,

$$d(\overline{AC}) = d(\overline{BD})$$

$$\sqrt{(d-0)^2 + (c-0)^2} = \sqrt{(b-a)^2 + (c-0)^2}$$

$$\sqrt{d^2 + c^2} = \sqrt{(b-a)^2 + c^2}$$

$$d^2 + c^2 = (b-a)^2 + c^2$$

$$d^2 = (b-a)^2$$

$$d = \pm(b-a)$$

$$d = b - a \quad \text{or} \quad d = a - b$$

Solving for b: $\quad b = d + a \quad$ or $\quad b = a - d$

Now, using $b = a - d$ and simplifying, we see that $d(\overline{AD}) = d(\overline{CB})$ so the trapezoid is isosceles.

$$d(\overline{AD}) = \sqrt{(b-0)^2 + (c-0)^2} = \sqrt{b^2 + c^2} = \sqrt{(a-d)^2 + c^2} = \sqrt{a^2 - 2ad + d^2 + c^2}$$

$$d(\overline{CB}) = \sqrt{(d-a)^2 + (c-0)^2} = \sqrt{(d-a)^2 + c^2} = \sqrt{d^2 - 2ad + a^2 + c^2}$$

14.

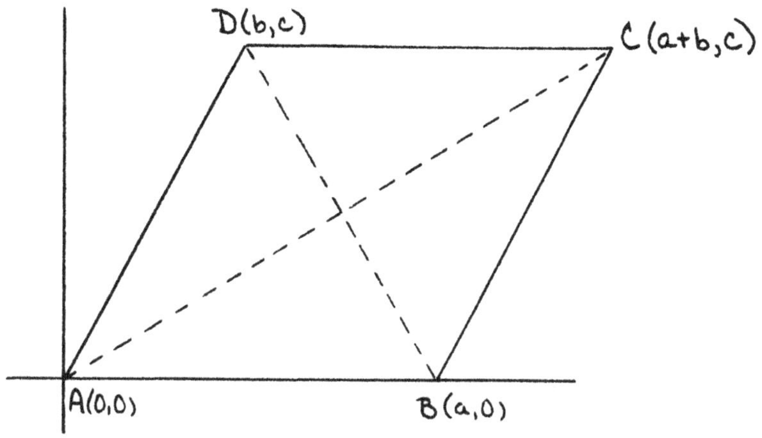

Let ABCD be a parallelogram placed in a coordinate system as shown. Since opposite sides of a parallelogram are parallel, we are justified using the same y-coordinate for points D and C. Since $\overline{AC} \perp \overline{BD}$, the slopes of their lines are negative reciprocals.

$$\text{slope of } \overline{AC} = -\frac{1}{\text{slope of } \overline{BD}}$$

$$\frac{c-0}{a+b-0} = -\frac{1}{\frac{c-0}{b-a}}$$

$$\frac{c}{a+b} = -\frac{b-a}{c}$$

$$c^2 = (a+b)(a-b)$$

$$c^2 = a^2 - b^2$$

Show the lengths of the four sides of the parallelogram are equal.

$$d(\overline{AB}) = a - 0 = a$$

$$d(\overline{BC}) = \sqrt{(a+b-a)^2 + (c-0)^2} = \sqrt{b^2+c^2} = \sqrt{a^2} = a \quad ; a > 0$$

$$d(\overline{CD}) = \sqrt{(a+b-b)^2 + (c-c)^2} = \sqrt{a^2} = a \quad a > 0$$

$$d(\overline{AD}) = \sqrt{(b-0)^2 + (c-0)^2} = \sqrt{b^2+c^2} = \sqrt{a^2} = a \quad a > 0$$

Since $d(\overline{AB}) = d(\overline{BC}) = d(\overline{CD}) = d(\overline{AD})$, the parallelogram is a rhombus.

20. Show the slope of $\overline{CP}$ is $\frac{2c}{2b - a}$, and that the equation of $\overline{CP}$ is $y = \frac{2c}{2b - a}\left(x - \frac{a}{2}\right)$.

$\overline{CP}$ passes through the points C(b, c) and $P\left(\frac{a}{2}, 0\right)$.

$$\text{slope of } \overline{CP} = \frac{c - 0}{b - \frac{a}{2}}$$

$$= \frac{c}{\frac{2b - a}{2}}$$

$$= \frac{c}{1} \cdot \frac{2}{2b - a}$$

$$= \frac{2c}{2b - a}$$

To find the equation of $\overline{CP}$, use the point-slope form of the equation of the line and let the point (x_1, y_1) be the point $P\left(\frac{a}{2}, 0\right)$:

$$y - y_1 = m(x - x_1)$$

$$y - 0 = \frac{2c}{2b - a}\left(x - \frac{a}{2}\right)$$

$$y = \frac{2c}{2b - a}\left(x - \frac{a}{2}\right)$$

Made in the USA
Monee, IL
05 November 2026

67807957R10109